# Writing in the Biological Sciences

## A COMPREHENSIVE RESOURCE FOR SCIENTIFIC COMMUNICATION

**Third Edition**

Angelika H. Hofmann
*Yale University*

New York Oxford
OXFORD UNIVERSITY PRESS

This book is dedicated to the memory of
***Francisco Javier Triana Alonso,***
who was and always will be larger than life to me

Oxford University Press is a department of the University of Oxford.
It furthers the University's objective of excellence in research,
scholarship, and education by publishing worldwide.
Oxford is a registered trademark of Oxford University Press
in the UK and certain other countries.

Published in the United States of America by Oxford University Press
198 Madison Avenue, New York, NY 10016, United States of America.


For titles covered by Section 112 of the US Higher Education
Opportunity Act, please visit www.oup.com/us/he for the latest
information about pricing and alternate formats.



Library of Congress Cataloging-in-Publication Data

Names: Hofmann, Angelika H., 1965- author.
Title: Writing in the biological sciences : a comprehensive resource for
scientific communication / Angelika H. Hofmann, Ph.D., Yale University.
Description: Third edition. | New York : Oxford University Press, [2019] |
Includes bibliographical references and index.
Identifiers: LCCN 2017052628 | ISBN 9780190852191 (pbk. : alk. paper)
Subjects: LCSH: Biology—Authorship. | Communication in science.
Classification: LCC QH304 .H64 2019 | DDC 570.1/5—dc23
LC record available at https://lccn.loc.gov/2017052628

9 8 7 6 5 4 3 2 1

Printed by LSC Communications, Inc., United States of America

# BRIEF CONTENTS

# CONTENTS

# FOREWORD

Academic science has a language of its own. Written and spoken by its practitioners, this language is sometimes practically unintelligible to the uninitiated. As a college or graduate-level student beginning to write scientific literature with highly technical content—papers, abstracts, proposals—your biggest challenge is to go beyond learning the jargon to convey complex factual information clearly and logically. As an instructor, your challenge is to teach your students this language through classroom practices. As a researcher, your challenge is to be fluent in the scientific language, producing peer-reviewed grants, manuscripts, and talks that may even reshape the language itself.

Despite years of training, in my experience, most new faculty researchers in science and medical fields are shocked and beleaguered by the writing workload required to build and sustain an independent research group. While they are scientifically ready for their new roles, much less attention has been paid to preparing their communication abilities. Yet, successful faculty researchers are also those with the communication experience and ability to advocate for their science effectively—especially as the pool of funds available for research continues to decline. To help prepare trainees for the challenges that lie ahead, the importance of scientific communication must be emphasized during college and graduate school. Moreover, as students practiced in scientific writing and communication also develop the ability to think logically and critically about the subject matter, these skills can be viewed as tools with which to train the budding scientific mind. Thus, the principles of sound scientific writing should be viewed as foundational and should be integrated throughout the curriculum of higher education much as chemistry is for those majoring in biology.

Why is the craft of scientific communication so often pushed to the side? This essential discipline requires specific educational resources to guide instructor and student. This text, *Writing in the Biological Sciences: A Comprehensive Resource for Scientific Communication*, is an invaluable contribution to the field. Through an accessible, clear, and concise, yet thorough, treatment of the subject, it instructs in the unique language of scientific communication. The book's practical approach of breaking down the most relevant forms of writing and presentation into their component parts enables the student to internalize key common principles that are discussed throughout the text. The text also demonstrates that those adept in scientific communication in the classroom—be it through lab reports, term papers, or in-class oral presentations—are those well on their way to crafting independent research publications, review articles, and seminar presentations.

Much like a box of paints, the techniques and information in this book are your toolkit with which you will communicate your own ideas and contributions to the outside world. You may be opening this book for one of many reasons. Perhaps you are using it as a required textbook. Perhaps you are coming to the formalized study of scientific writing for the first time, or perhaps you are someone who is seeking specifically to improve your scientific writing skill. Keep in mind that as with a foreign language, true proficiency in scientific communication can take years to achieve. Hold on to this volume in the years to come, because you will refer to the guidelines within again and again.

Tammy Wu, PhD
Yale University

# PREFACE

Communication plays a fundamental role in the sciences, and beyond. It is the engine that propels virtually all progress. Without good communication skills, those in science stand little chance of publishing their work, obtaining funds, attracting a wide audience when giving a talk, or getting a new job. In the sciences, even the most promising discovery means little if it cannot be communicated successfully. In fact, it is more often the case that advancements are limited not by their technical merit but by ineffective articulation. Thus, clear communication is a necessity, not an option, for scientists and students in this field.

*Writing in the Biological Sciences: A Comprehensive Resource for Scientific Communication*, third edition, serves as an all-inclusive "one-stop" reference guide to scientific writing and communication for budding professionals in the life sciences and other fields. The book is intended as a free-standing textbook for a corresponding course on writing in the sciences as well as an accompanying, practical text or reference guide in courses with writing-intensive components. It covers all the basics of scientific communication that students need to know and master for successful careers. The book lays the foundations for future professional writing by starting with basic scientific writing rules and principles and applying these to composing lab reports, literature reviews, summaries, and critiques and eventually to full-fledged scientific research articles, review articles, and grant proposals. Thus, for example, lab reports that follow the general format provided in this book provide the groundwork for later scientific research articles that get published in academic journals; term papers/literature reviews are the basis for review articles, and research proposals for grant proposals. Practical advice for organizing academic presentations and posters as well as for putting together job applications is also included.

Through extensive and annotated practical examples and easy-to-understand and easy-to-follow rules and guidelines, this book explains how to write clearly as an author in the sciences, and how to recognize shortcomings in one's own writing, as well as in that of others. It does so not only by providing crucial knowledge about the structure and delivery of written material but also by explaining how readers go about reading. Furthermore, potential problem areas in written and oral presentations are pointed out, and many examples are provided for wording certain sections in research papers, grant proposals, or scientific talks.

There are numerous hallmark features of this text, including:

**Practical organization.** As a result of extensive class testing, the text has been revised repeatedly to reflect the interests, concerns, and problems undergraduate students encounter when learning how to communicate in their disciplines. The table of contents is divided into four distinctive parts. Parts I and II follow a logical progression from the basics of scientific writing style and composition to constructing effective figures and tables and selecting references. Part III applies these basic rules and guidelines to planning and organizing foundational precursors of scientific publications, namely lab reports, literature reviews, summaries, and critiques, to the specifics of how to write each major section of a scientific research paper and review article. Part IV covers more advanced scientific communication, including how to compose grant proposals, posters, oral presentations, and job applications.

**Comprehensive coverage.** The text includes detailed discussions of the main scientific documents undergraduate students in the sciences encounter, including laboratory reports, scientific research papers, literature reviews, review articles, proposals, oral presentations, posters, and job applications. The broad coverage of multiple forms of communication allows the text to serve as a writing guide for science writing courses as well as a companion text for advanced biology courses that have a writing-intensive component or for work to be published by a student. In addition, the text provides comprehensive coverage of writing mechanics, style, and composition.

**Numerous real-world and relevant examples.** The in-chapter examples are derived directly from real lab reports, scientific research papers, review articles, and grant proposals in the biological sciences. Throughout the chapters, common pitfall examples are followed by successful revisions, and annotated examples provide explanations of various text elements and concepts. These annotated examples bring to life the rules and guidelines presented throughout the chapters.

**Extensive exercise sets and end-of-chapter summaries.** Chapter summaries reference the most important concepts in an easy-to-understand format. The end-of-chapter exercises and problems review style and composition guidelines and encourage students to apply the presented rules and guidelines to their own writing. Answers to the exercises are provided in a separate appendix.

**Writing guidelines and checklists for revisions.** Straightforward rules and guidelines presented in the book provide the basis for writing scientific articles, proposals, and job applications and for creating clear posters and oral presentations. Explanations of these basic writing rules and guidelines are followed by common pitfall examples, as well as by suggestions and advice to revise one's work successfully. Checklists at the end of each chapter aid readers in remembering and applying these rules when writing or revising a document.

**Sample wording for scientific documents and presentations.** Beginning scientific writers will especially value the many tables with sample sentences that apply to different sections of a scientific research article, review article, or grant proposal. Anyone presenting data at meetings and conferences will find the sample phrases and advice on creating and delivering a talk or poster highly useful.

*Writing in the Biological Sciences: A Comprehensive Resource for Scientific Communication* teaches students how to practice writing and thinking in the sciences and beyond. It lets them learn by example from the writings of others. Familiarity with the nuances of these elements will be enhanced as students read scientific literature and pay attention to how professional scientists and others write about their work. Students will see improvement in their own writing skills by understanding and repeatedly practicing reading, writing, and critiquing of others' work. Writing a clear research article or grant proposal or presenting an articulate talk can be difficult for any scientist, but this difficulty is by no means insurmountable. Ultimately, with guidance and practice, students should be able to write papers or proposals that sparkle with clarity and deliver engaging presentations. As they write their own papers or prepare their own posters, they will recognize that every project has its unique challenges and that they will need practice and good judgment to apply all the writing and communication guidelines presented herein. In giving due attention to composition, style, and impact, students will improve their communication skills significantly, and this book will have accomplished its purpose.

## NEW TO THE THIRD EDITION

Since the publication of the first and second editions, I have heard from many professors and students that they found the text's comprehensive, practical, and hands-on approach to be of great value as they produced a wide range of scientific documents. Listening to their comments, I revised the text with the goal of expanding these hallmark features and providing additional new resources. Specific updates and improvements in the third edition include the following:

- **New sections on the scientific method and scientific writing** provide context and understanding of the importance and approach of scientific communication.
- **A new section on scientific ethics** discusses the importance of these issues and provides guidance on key questions.
- **A new section on basic statistical analysis** explains the fundamentals of reporting statistical data and analyses in a scientific context.
- **An expanded section on plagiarism** guides students on avoiding this pitfall and makes them aware of important bioethical issues.

- **A glossary of scientific and technical terms** confers a convenient place to look up specific terminology used in scientific communication.
- **A new section on the most common interview questions** allows those seeking jobs to prepare for being interviewed.
- **An updated layout of the text and chapter overviews** enables readers to find information quickly throughout the book.
- **Updated PowerPoint slides** accompany the revisions of the third edition of *Writing in the Biological Sciences: A Comprehensive Resource for Scientific Communication.*

## ACKNOWLEDGMENTS

This book includes many ideas and "specimens" that I, as a scientist, instructor, and editor, have collected over the years. A few of these "specimens" are originals. Many are a variation of someone else's original. Others are cited intact from their respective sources. Without these "specimens" and samples, this book would not have been possible. For their contribution, I would therefore like to especially acknowledge my students, friends, and colleagues from the Chinese Academy of Medical Sciences, Max Planck Institute, Fritz-Haber Institute, Humboldt State University, Karolinska Institute, University of Carabobo, University of Pittsburgh, University of Massachusetts at Worcester, Washington University, and Yale University who have shared information and ideas across the sciences. I am particularly thankful to all those who were courageous enough to allow me to use draft sentences, paragraphs, or sections as examples or problems in this book as well as to those providing me with extensive and very specific samples: Maxx Amendola, Irene Bosch, Mark Bradford, Jaclyn Brown, Stephane Budel, Philip Duffy, Mónica I. Feliú-Mójer, Nikolas Franceschi Hofmann, Alison Galvani, Roland Geerken, Jun Korenaga, Annie Little, Amanda Miller, Klaus von Schwarzenberg, Jeffrey Townsend, and Tammy Wu. Without these samples the book would not be nearly as effective in exemplifying clear writing.

I am also grateful to all the reviewers who have edited and commented on various draft chapters, including:

Andrea Aspbury
*Texas State University*

Scott D. Banville
*Nicholls State University*

Lynn Carpenter
*University of Michigan*

Mary Carla Curran
*Savannah State University*

Keith Gibbs
*Tennessee Technological University*

Douglas Glazier
*Juniata College*

Christopher J. Grant
*Juniata College*

Andrea Henle
*Carthage College*

Brooke Jude
*Bard College*

Michelle Kulp McEliece
*Gwynedd Mercy University*

Alan McGreevy
*University of Winnipeg*

Jennifer Metzler
*Ball State University*

Roman J. Miller
*Eastern Mennonite University*

Lindsay Emory Moore
*University of North Texas*

Judith D. Ochrietor
*University of North Florida*

T. Page Owen
*Connecticut College*

Nikki Panter
*Tennessee Technological University*

Marc Perkins
*Orange Coast College*

Alice Jo Rainville
*Eastern Michigan University*

Cara Shillington
*Eastern Michigan University*

Jeffrey H. Toney
*Kean University*

Charlotte M. Vines
*University of Texas El Paso*

Neal J. Voelz
*St. Cloud State University*

Larissa Williams
*Bates College*

Jinchuan Xing
*Rutgers University*

I would also like to acknowledge those reviewers of the first and second editions, whose advice and comments were instrumental in establishing the foundation of this text: Daniel Abel, Andrea Aspbury, Robert Benard, Daniel G. Blackburn, Lisa Ann Blankinship, Christopher P. Bloch, Chad E. Brassil, Heather Bruns, Arthur Buikema, Alyssa C. Bumbaugh, Douglas J. Burks, Reneé E. Carleton, Lynn L. Carpenter, Dale Casamatta, Deborah Cato, David T. Champlin, Kendra Cipollini, Francisco Cruz, Charles Elzinga, Robert Feissner, Kirsten Fisher, Laurel Fox, Christopher J. Grant, Blaine Griffen, Carl James Grindley, Glenn Harris, Christiane Healey, Leif Hembre, Evelyn N. Hiatt, W. Wyatt Hoback, Terry Keiser, Lani Keller, Tali Lee, Michelle Mabry, Joshua Mackie, Nusrat Malik, Jesse M. Meik, Jennifer A. Metzler, Daniel Moon, Barbara Musolf, Judith D. Ochrietor, Michael O'Donnell, T. Page Owen, Jr., Helen Piontkivska, Mary Poffenroth, Byrn Booth Quimby, Ann M. Ray, Letitia M. Reihcart, Ann E. Rushing, Allen Sanborn, Roxann Schroeder, Robert S. Stelzer, Ken G. Sweat, Katerina Thompson, Charlotte Vines, Neal J. Voelz, and Nancy Wheat.

Particular thanks go to Betty Liu, Fiona Bradford, Paola Crucitti, Riccardo Missich, Rudolf Lurz, Tammy Wu, Francois Franceschi, Jennifer Powell, Stephane Budel, and John Alvaro for their encouragement as well for as their critical comments and the many helpful discussions over the years. My deepest gratitude is in memory of Francisco Triana Alonso for his unwavering belief, support, and pride in me and this endeavor—beginning with my first sprouting ideas, to the delivery of them to students, and ultimately, to the publication of this book.

Finally, I would like to express appreciation to everyone at Oxford University Press: Jason Noe, senior editor; Andrew Heaton, associate editor; Nina Rodríguez-Marty, editorial assistant; Patrick Lynch, editorial director; John Challice, publisher and vice president; Bill Marting, national sales manager; Frank Mortimer, director of marketing; Tina Chapman, marketing manager; Colleen Rowe, marketing assistant; Lisa Grzan, production manager; Denise Phillip Grant, production team leader; Patricia Berube, project manager; Michele Laseau, art director; and Todd Williams, designer.

# PART I

# Scientific Writing Basics

CHAPTER 1

# Science and Scientific Communication

THIS CHAPTER PROVIDES AN OVERVIEW OF:

- The importance of scientific communication
- What writers should be aware of to communicate effectively and clearly to readers
- The overall approach and structure of this book

## 1.1 THE SCIENTIFIC METHOD AND COMMUNICATION

### ➤ Understand the scientific method

Science studies the natural world. It includes many different fields from the life sciences to earth sciences to physical sciences. Through research studies and experiments, scientists try to glean information and derive models to explain diverse phenomena. The series of steps involved in such studies have a similar pattern throughout the diverse scientific fields. These steps are collectively known as the scientific method and consist of:

- Asking a question/making an observation
- Proposing a hypothesis
- Testing the hypothesis through experiments
- Analyzing and interpreting the data/drawing conclusions
- Communicating results

A scientific hypothesis is a highly probable proposition, based on observations and prior knowledge that can be investigated further. If the hypothesis is wrong, scientists often come up with a new hypothesis. They then repeat the scientific steps to test it by experimentation or observation and publish their findings. If the hypothesis is correct and broad enough in scope, and if other scientists agree with it, it can turn into a theory and eventually into a law, on which scientists can build to advance research and knowledge further.

It is the last point of the scientific method, communicating results, with which this book is mainly concerned. It focuses primarily on the biological sciences, given that the majority of students are taking classes or majoring in these scientific fields, and that these fields require much writing. However, the underlying principles presented here can certainly also be applied to other scientific disciplines.

Regardless of the field, without clear communication, even the best results in science mean little. As a science student, it is therefore essential to be trained not only in observation, formulation of a hypothesis, and experimentation, but also in scientific communication. Learning how to communicate in science will prepare you well for the professional world. Without these skills, it will be difficult to succeed as your career will depend on publications, funding for research will require successful proposals, and to become known in your field necessitates presentations and posters. Communication in the professional scientific world includes:

- Original scientific research articles—to communicate findings to other scientists and to the public
- Review articles—to glean and communicate in-depth interpretations of current topics published in research articles
- Grant proposals—to apply for funding of research
- Posters and oral presentations—to present your work visually and orally
- Science news articles, blogs, or lectures—to communicate science effectively to the public, students, and others
- Evaluation of the work of others—as a reviewer of a manuscript, for example
- Letters of recommendation
- Progress reports
- Cover letters
- Job applications

This book is meant to prepare you for these professional skills by teaching you how to research and use references, present and report on data, write lab reports (the precursors for research articles), compose summaries and critiques (to learn how to write reviews and how to review papers), draft grant proposals, and prepare presentations, posters, and even job applications. To produce these communications successfully, you need a good foundation, including basic writing skills starting at the smallest units (words) and working up to the larger ones (note taking, paragraphs, articles, reports).

### ➤ Learn how to write in science

Poor style and composition in scientific writing is often due to a lack of training and the difficulty of training. The majority of writers are unaware of how to identify words, sentences, or paragraphs that may pose problems for readers. Most are aware of certain "rules" taught to them in high school English composition classes but not of the fine-tuned principles and guidelines that would benefit them as college students in the sciences or as professionals.

To be a successful professional scientific writer, basic English composition is not enough. You need to know what information readers in the field expect to find and where in the text they expect to find it. You need to understand the style and composition expected in scientific articles, proposals, and presentations. You also need to realize that all too often you are too "close" to your own writing to judge it and therefore need to work as a team editing and revising your work repeatedly.

### ➤ Practice writing

Use the rules and guidelines in this book to help you grasp fully the necessary format expected for the respective scientific documents. Realize, though, that reading and hearing about these rules and guidelines is not enough. You, as the writer, must *practice* writing and thinking within this structure and must learn by example from the writings of others. Familiarity with the nuances of these elements will be enhanced as you read scientific literature and pay attention to how professional scientists write about their work. Moreover, note that practicing writing typically includes writing multiple drafts, asking your instructor and peers for feedback, and establishing a practice of reading your text over to help identify awkward phrasing or lack of clarity in your document. Mastering the skill of scientific writing and communication, like all others, takes time—usually years. In giving due attention to composition, style, and impact, your communication skills will receive the necessary foundation and evolve into those needed for professional writing, and this book will have accomplished its purpose.

## 1.2 NECESSARY TECHNICAL SKILLS

### ➤ Acquire the necessary technical skills for science communication

To compose scientific documents and presentations, technical expertise is also needed. This expertise encompasses computer literacy and skills in specific computer programs, including:

- Microsoft Word or Apple Pages for writing and formatting documents
- Microsoft Excel or Apple Numbers for making tables and preparing graphs
- Statistical software packages (e.g., R, Minitab, or SigmaPlot) for statistical analysis
- Microsoft PowerPoint or Apple Keynote for creating slides and posters for presentations

- Google Docs for composing and editing working documents among authors
- Google Sheets to create and edit spreadsheets with others online

If you find yourself unfamiliar with any of these programs, consider taking some classes or completing an online tutorial (see also Sections 5.7, 6.6, and 13.5). It is essential to have these basic skill sets when it comes to scientific writing.

## 1.3 SCIENTIFIC WRITING VERSUS SCIENCE WRITING

**➤ Distinguish between scientific writing and science writing**

People often use the terms "scientific writing" and "science writing" synonymously, but they are actually quite different. Scientific writing is a form of technical writing by scientists for other scientists. Unlike other styles of writing, this style of writing is not meant to be creative but fact-based and objective. It reports on studies, observations, and findings in a specific format. For example, peer-reviewed journal articles, which report on primary research and are published in scientific journals such as *Science* and *Nature*, fall into this category. Grant proposals, which seek funding, and literature review articles, which summarize and interpret published research, also belong to this genre.

---

**Example 1-1** Beta-thalassemia, an inherited morbid hemoglobinopathy, is caused by single point mutations in the gene for beta-globin. Our long-term goal is to diagnose and treat hemoglobinopathies by using high fidelity editing reagents, such as peptide nucleic acids, carried in targetable nanoparticles. To date, our group has optimized editing reagents using single stranded DNA to correct the point mutation that causes beta-thalassemia. . . .

---

Science writing, on the other hand, is nontechnical writing about science for a general audience. It is the form of journalism you find in *Scientific American*, *National Geographic*, or in articles on science published by the *New York Times*. Science writing is not a how-to manual or a review paper or research article. Science writers explain important and interesting topics in science and technology and lay out the broader social effects of these topics to a wide public audience. Science writers use nontechnical language to explain and interpret scientific concepts or processes in a way that is clear to a lay audience.

---

**Example 1-2** Beta-thalassemia is an often fatal, inherited red blood cell disease, which is caused by a genetic mutation in the beta-globin gene. Scientists plan to repair mutations causing the disease by excising the mutated DNA and replacing it with DNA pieces containing the correct sequence. The correct DNA pieces will be carried into the cells on tiny particles the size of a few nanometers, which are able to cross the cell membrane. . . .

---

## 1.4 SCIENCE COMMUNICATION AND ETHICS

### ➤ Be ethical

Good science does not only need to be communicated well and effectively, it also has to pass scientific ethics. Ethical norms cover a wide range of topics, from use of human subjects, such as fetal tissue, to fraud, sponsorship of research, and plagiarism (see also Chapter 4, Section 4.6). These norms are governed by standards that are important for the following reasons:

- To ensure accuracy and truth
- To ensure mutual respect, fairness, and trust
- To hold scientists accountable

Accordingly, codes of ethical conduct for scientists have been established by diverse professional organizations, journals, universities, institutes, and government agencies. Key agreements include the Declaration of Helsinki, a set of ethical principles for research on human subjects developed by the World Medical Association (available at www.wma.net) and the Nuremberg Code, a set of ethical principles for experimentation on humans established after World War II (http://ohsr.od.nih.gov/guidelines/nuremberg.html).

Ethics in science also extends to scientific writing and communication. Scientific misconduct is federally defined as intended actions, such as fabrication and falsification of data and plagiarism. Nearly all researchers view these as unethical. Note, though, that human errors, sloppiness, miscalculations, bias, disparities of methods and interpretations, and even negligence are not classified as misconduct. Other deviations not defined as misconduct also are viewed as unethical by most scientists: stealing someone else's ideas or data, submitting or publishing the same papers in different journals, including someone who has not contributed to a project as an author on a paper, filing a patent without informing collaborators, asking sexual favors in exchange for authorship or a grade, exploiting students and postdocs, misrepresenting facts on a CV or other document, and more.

### ➤ Aid in distinguishing facts and fake news

At a time when sharing and accessing information is easier than ever and when fake news has been on the rise, clear and high quality scientific communication has never been more important. The more clearly and effectively information is laid out, the easier it is understood and the less likely it can be misinterpreted and misreported. Not only must data and results be stated clearly, they also need to be verifiable and supported by evidence. Thus, support your statements with evidence—through data, statistical analysis, or peer-reviewed publications.

---

**Example 1-3** X has been shown *in vitro* to mediate gene correction in human blood cells **[4]**.

---

**Example 1-4** HSCs of the fetal liver express high levels of c-kit receptor **(Figure 2)**.

Verify information when you compose your documents rather than relying on references in other articles, subjective opinions, policy-based evidence, unsupported facts, and non-peer-reviewed reports or Web sources (see also Chapter 4, Section 4.2). Do not report your opinion as fact. Rather, clearly identify your opinion and conclusions as such (*Our findings indicate that . . . ; A possible model of X could be . . .*).

Valuing and respecting the scientific method and the peer-review process is essential when it comes to drafting, publishing, and reviewing manuscripts. I highly recommend publishing in respected, peer-review journals, also known as primary sources, rather than on non-peer-reviewed websites. Find out what the most respected journals are in your field—inquire with your professors and librarians and see Chapter 4.

If you are unsure about information you receive or find, make use of the scientific method: gather evidence, check sources, deduce, hypothesize and synthesize conclusions yourself rather than relying on those of others. Knowing and using these practices will aid you and the general public in distinguishing facts from fake news.

## 1.5 READERS AND WRITERS

### ➤ Understand how readers go about reading

To write clearly in the sciences, you not only need professional writing tools, such as rules of style, composition, and format, but you also need to understand what readers expect to find and how they go about reading. Expectations and perceptions of readers have been widely studied in the fields of rhetoric, linguistics, and cognitive psychology. One of the main findings is that readers do not just read. Instead, they immediately interpret what they read.

Let me illustrate how readers go about reading using the example of one word:

**Water!**

Immediately, on reading this single word, you will have a picture in your mind. Some of you will think of water as in a dangerous flood or tsunami, others will think of going swimming on a hot day, yet others will think of getting water to drink when they are thirsty or the excitement they feel on finding water for which they have been drilling. In other words, different readers interpret this single word differently.

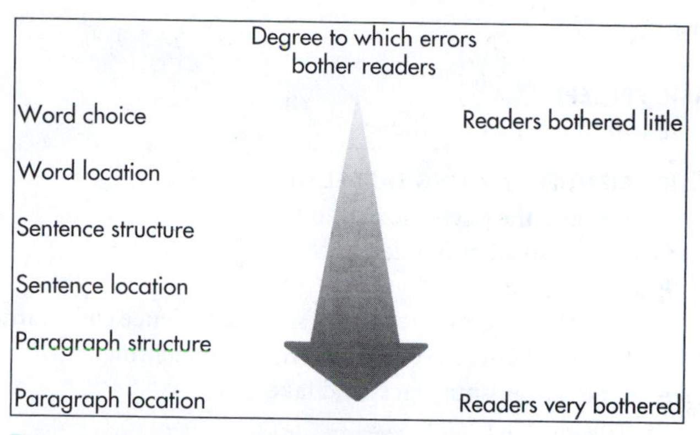


Figure 1.1 Degree to which errors in writing bother readers.

In science, textual interpretations are based not only on words, but also on sentences and paragraphs, and above all, on the organization of these elements. In fact, it is the larger structures, such as paragraphs and their locations, with which the reader is much more concerned than the smaller elements when it comes to understanding what has been written in the sciences (see Figure 1.1). As an author, you need to be conscious that the correlation of structure and function in a sentence, paragraph, or section is what underlies the science of scientific writing. It is therefore of utmost importance that you understand readers.

Aside from these elements of text composition, communication in science also depends on effective graphics depicting data or communicating complex biological phenomena to complement the writing. These visual elements are often key to providing the required evidence to convince readers of your data interpretations and to provide pictorial models that readers can remember. Understanding the needs of your audience in that respect is thus also important in becoming a successful writer.

## ➤ Write with the reader in mind

In the professional world, success in writing is determined by whether your readers understand what you are trying to say. You need to write clearly so that readers can follow your thinking and so that you achieve the greatest possible impact. To "write with the reader in mind" means to consider how the reader interprets what you have written. It requires you to construct your writing clearly, concisely, and at the right level, so that the reader can follow and understand what you want to say immediately. This rule should be viewed as the ***Basic Precept*** around which all other writing rules revolve.

## SUMMARY

### BASIC PRECEPT

Write with the reader in mind.

### BASIC SCIENTIFIC WRITING GUIDELINES

1. Understand the scientific method.
2. Learn how to write in science.
3. Practice writing.
4. Acquire the necessary technical skills for science communication.
5. Distinguish between science writing and scientific writing.
6. Aid in distinguishing facts and fake news.
7. Be ethical.
8. Understand how readers go about reading.

CHAPTER 2

# Fundamentals of Scientific Writing—Style

IN THIS CHAPTER YOU WILL FIND:

- Basic scientific writing rules for words and word location
- General advice on style in science
- Basics for composing technical sentences
- Information on technical style (including first and third person, active and passive voice, past and present tense, sentence length, verbs and action, lists and comparisons, and common errors)

Basic principles of scientific style and composition underlie all professional scientific communication. Yet, these writing rules are often not taught or refined sufficiently in the sciences. This chapter groups essential scientific writing rules and guidelines for the elements of style and composition into three groupings: individual words, word location, and sentence structure and technical style.

## 2.1 INDIVIDUAL WORDS

### ➤ Use precise words

Words in science should be precise. Imprecise word choices can be problematic because they will be unclear to readers. Such word choices should be revised. Following are some examples:

| | | |
|---|---|---|
| 👎 | **Example 2-1** | Plants were kept in the cold overnight. |
| 👍 | **Revised Example 2-1** | Plants were kept **at 0°C** overnight. |
| 👎 | **Example 2-2** | Reagent Y was mixed with X. |
| 👍 | **Revised Example 2-2** | Reagent Y and X were mixed **together**.<br>OR<br>Reagent Y was mixed **using** X. |

### ➤ Use simple words

Scientific writing has many technical terms that are difficult to understand because they are intended for a very specific audience. To keep your writing from being too heavy, choose simple words for the rest of the sentence. Such use will also ensure that as many readers as possible (including non-native speakers) can understand what has been written.

The following revised sentences are much easier for readers to understand because their word choices are much simpler:

| | | |
|---|---|---|
| 👎 | **Example 2-3** | We utilized UV light to induce mutations in *Arabidopsis*. |
| 👍 | **Revised Example 2-3** | We **used** UV light to induce mutations in *Arabidopsis*. |
| 👎 | **Example 2-4** | For the purpose of examining cell migration, we dissected mouse brains. |
| 👍 | **Revised Example 2-4** | **To** examine cell migration, we dissected mouse brains. |

### ➤ Omit unnecessary words and phrases

Your writing should be as brief as possible but clear. Many sentences in science appear complex because they contain unnecessary words and phrases or redundancies.

## Unnecessary Words

**Example 2-5** The sample size was not quite sufficiently large enough.

**Revised Example 2-5** The sample size was not **large** enough.

The following words usually can and should be omitted entirely:

| | | | | | |
|---|---|---|---|---|---|
| actually | basically | essentially | fairly | much | really |
| practically | quite | rather | several | very | virtually |

In the following examples of redundancies, all the words in parentheses can be omitted:

| | |
|---|---|
| (already) existing | at (the) present (time) |
| blue (in color) | cold (temperature) |
| (completely) eliminate | (currently) underway |
| each (individual) | each and every (choose one) |
| (end) result | estimated (roughly) at |
| (final) outcome | first (and foremost) |
| (future) plans | never (before) |
| period (of time) | (true) facts |

## Unnecessary Phrases

Aside from avoiding unnecessary words, omit unnecessary phrases to make your writing shorter and clearer.

**Example 2-6** It is well known that there are three subtypes of the KL-2 virus.

**Revised Example 2-6** There are three subtypes of the KL-2 virus.

**Example 2-7** It has been reported that the population of Alaskan sea otters has been declining.

**Revised Example 2-7** The population of Alaskan sea otters has been declining (Lopez et al., 1995).

Commonly used unnecessary phrases that can usually be deleted include:

| | |
|---|---|
| there are many papers stating . . . | it is speculated that . . . |
| it was shown to . . . | it has been found that . . . |
| it has long been known that . . . | it has been reported that . . . |

### Phrases to Avoid

| Avoid | Better |
|---|---|
| a considerable number of | many |
| a consequence of | because |
| at no time | never |
| based on the fact that | because |
| despite the fact that | although |
| due to the fact that | due to |
| during the course of | during, while |
| during the time that | while, when |
| elucidate | explain |
| employ | use |
| facilitate | enable |
| first of all | first |
| for the purpose of | to |
| has the capability of | can, is able |
| in case | if |
| in many cases | often |
| in order to | to |
| in some cases | sometimes |
| in the absence of | without |
| in the event that | if |
| it is of interest to note that | note that |
| majority of | most |
| no later than | by |
| on the basis of | by |
| prior to | before |
| referred to as | called |
| regardless of the fact that | even though |
| so as to | to |
| terminate | end |
| the great majority of | most |

### ➤ Avoid too many abbreviations

In these days of texting and social media, the use of acronyms has become more and more common. Too many abbreviations can be confusing to the reader, however. Thus, abbreviations or acronyms should be kept to a minimum. Similarly, nonstandard abbreviations need to be limited; otherwise, the reader will get lost.

**Example 2-8** The increase in SSTIs caused by CA-MRSA but not by MSSA is especially marked in the pediatric population as a study from PCH shows.

The previous example may be perfectly intelligible to expert colleagues but will be unintelligible to others.

**Revised Example 2-8** The increase in **soft tissue infections** caused by **community-acquired methicillin-resistant** *Staphylococcus aureus* but not by **methicillin-sensitive** *S. aureus* is especially marked in the pediatric population as a study from **Philadelphia Children's Hospital** shows.

If lengthy terms are used often throughout a scientific document, then abbreviations that have been introduced can be used instead. In these cases, always define acronyms—for example, *ultraviolet (UV)* or *nuclear magnetic resonance (NMR)*—and essential nonstandard abbreviations on first appearance, although not in a title or abstract.

**Example 2-9** Over the past two decades, **methicillin-resistant** *Staphylococcus aureus* **(MRSA)** has frequently been isolated from hospitalized patients, but also in the community.

## ➤ Use correct nomenclature and terminology

### General Scientific Nomenclature

Use correct vocabulary, nomenclature, and terminology to avoid being misunderstood or confusing the reader. If you are not sure, do not guess. Take the time to look up terms in a dictionary or other reference book. Dictionaries for the biological, medical, and other scientific fields, as well as online dictionaries, are listed in the back of this book.

Commonly used scientific nomenclatures include:

**Species and most Latin derivates** are in *italics* (*in vivo*, *Physcomitrella patens*)
**Human genes**: all caps and *italics* (*ADH3, HBA1*)
**Mouse genes**: first letter capitalized, the rest lower case, *italics* (*Sta, Shh, Glra1*)
**Human proteins**: capitals, no italics (ADH3, HBA1)
**Mouse proteins**: like genes, but no italics (Sta, Shh, Glra1)
**Restriction enzymes**: a combination of *italics* and nonitalics (e.g., *Bam*HI); check with the supplier for the correct nomenclature if unsure

### Misused and Confused Terms

Words are not always what they seem. Quite a few words and expressions in science are commonly misused and confused. Watch out for these misused and confused terms. Use a dictionary when you write if needed or look up definitions on the Internet. Table 2.1 briefly lists the most commonly confused terms. A more comprehensive list can be found in Appendix A.

**TABLE 2.1 List of Most Commonly Confused Terms**

| | | |
|---|---|---|
| affect | verb, meaning "to act on" or "to influence" (note: affect is often also used as a noun, meaning emotion or desire, in psychology) | The addition of KI-3 to MZ1 cells **affected** their growth rate (i.e., it could have increased or decreased or induced). |
| effect | 1. noun, meaning a result or resultant condition<br>2. verb, meaning "to cause" or "to bring about" | 1. We examined the **effect** of KI-3 on MZ1 cells.<br>2. The addition of NaCl to our solution **effected** slower cell growth (i.e., it caused or brought about). |
| as | conjunction, used before phrases and clauses | Let me speak to you **as** a father (= I am your father and I am speaking to you in that character)<br>**as** we just mentioned<br>**as** described previously, **as** described by . . . |
| like | preposition, meaning "in the same way as" | Let me speak to you like a father (= I am not your father but I am speaking to you as your father might).<br>Our observations were **like** those of Andrews et al. |
| which | use with commas for nondefining (nonessential) clauses | Dogs, **which** are cute, recovered. |
| that | use without commas for essential clauses; if the section of a sentence introduced by "that" is omitted, the meaning of the sentence is changed or may not be apparent | Dogs **that** were treated with antidote recovered. |
| contrary to | preposition meaning "in opposition to" | **Contrary to** our expectations, addition of $Mg^{++}$ did not alter our results. |
| on the contrary | usually used only in spoken English; used when one says a statement is not true | "It's exciting!" "**On the contrary**, it's boring!" |
| on the other hand | rarely used in scientific English; used when adding a new and different fact to a statement | On the one hand, plants need sunshine, and **on the other hand**, they need rain. |
| in contrast | used for two different facts that are both true, but pointing out the surprising difference between them | pH values increased for prokaryotes. **In contrast**, no pH difference was observed in eukaryotes. |

## 2.2 WORD LOCATION

### ➤ Establish importance

Word location within a sentence can help authors to guide and influence readers. The format and structure authors use to present information will lead the reader to interpret it as important or less important. To decide on the best placement of words within a sentence, it is crucial that authors establish importance (i.e., decide which information is most important and should therefore be emphasized). In general, the end position in a sentence is more emphasized than the beginning position, and the main clause is more emphasized than the dependent clause.

Consider the following two versions of a sentence:

**Example 2–10**

a Although vitamin B6 reduces the risk of macular degeneration, taking it has some side effects.

b Although taking vitamin B6 has some side effects, it reduces macular degeneration.

Because of the placement of information in these sentences, most readers view the overall recommendation of vitamin B6 as negative in sentence **a**, whereas that of sentence **b** is viewed by most as positive.

### ➤ Place old, familiar, and short information at the beginning of a sentence

Readers expect to see familiar information at the beginning of a sentence. This placement allows them to link information back to the topic or subtopic of a document. For example, in a section or paragraph about bees, readers expect to find "bees" mentioned at the beginning of a sentence.

**Example 2-11a** **Bees** belong to the genus *Apis.*

### ➤ Place new, complex, or long information at the end of a sentence

Readers anticipate new information at the end of a sentence (or paragraph) where it is more emphasized, as in the following example:

**Example 2-11b** Bees produce **honey**.

OR

Bees collect not only pollen but also **nectar**.

Readers also follow information more easily when short items are listed before long or complex ones in a series, because such listing is less disruptive to the flow of a sentence, as illustrated in the next example.

---

**Example 2-11c** Solitary bees build nests on twigs, in wood, in tunnels**, or between rocks in the ground**.

---

If information is placed where most readers expect to find it, it is interpreted more easily and more uniformly. Thoughtful placement of information becomes even more important in paragraphs where readers expect information to link sentences to each other. Skilled application allows the information at the end position of a sentence to be linked to that at the beginning, or topic position, of the next sentence. Use of such word location creates good "flow" of a paragraph, as in the following example:

---

**Example 2-11d**

***Macular degeneration*** is affected by ***diet. One of the diet components*** that influences the progression of macular

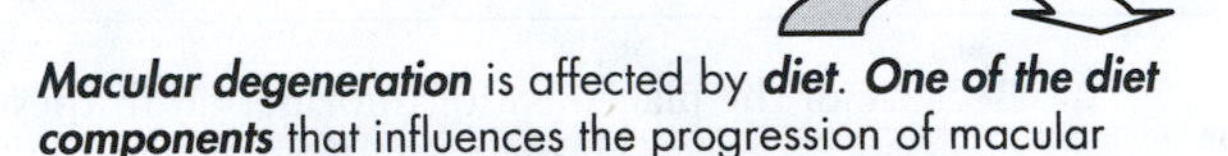

degeneration is ***vitamin B6***. Although ***vitamin B6*** seems to reduce the risk of macular degeneration, it may have some ***side effects***.

---

## ➤ Get to the subject of the main sentence quickly

Readers understand a sentence more easily if its subject is readily available. When a sentence is started with several words before its subject/topic, readers find it difficult to see what the sentence is about. Avoid such long introductory phrases and subjects.

---

**Example 2-12** **Because high-resolution EM methods can show fine details of oligomeric proteins or complexes such as ribosomes,** *these methods* may show where nucleic acid strands are located.

---

**Revised Example 2-12** ***High-resolution EM methods*** may show where nucleic acid strands are located because these methods can show fine details of oligomeric proteins or complexes such as ribosomes.

---

## ➤ Place the verb immediately after the subject and the object right after the verb

After reading the subject, readers expect to find the verb. If no verb follows the subject directly, readers will be looking for it. They will ignore interrupting material between subject and verb, especially if it is lengthy, considering it less important.

**Example 2-13** Rhinovirus, an Enterovirus belonging to the family *Picornaviridae*, which consists of 37 species grouped into 17 genera including pathogens such as Poliovirus, Saffold virus, Coxsackie A virus, and Hepatitis A virus, causes 10–40% of the common cold.

This sentence obstructs the reader, because the grammatical subject ("Rhinovirus") is separated from its verb ("caused") by thirty words.

Often, an interruption can be moved to the beginning or the end of a sentence, depending on whether it is connected to old or new information in the sentence. At other times, the author should consider splitting the information into two sentences or omitting it altogether if it is not essential.

**Revised Example 2-13a** **Rhinovirus is** an Enterovirus belonging to the family *Picornaviridae*, which consists of 37 species grouped into 17 genera including pathogens such as Poliovirus, Saffold virus, Coxsackie A virus, and Hepatitis A virus. **It causes** 10–40% of the common cold.

**Revised Example 2-13b** **Rhinovirus causes** 10–40% of the common cold.

Readers also like to get past the verb to the object of a sentence quickly. Thus, authors should avoid any interruptions between verb and object by placing interrupting passages either at the beginning or at the end of the sentence.

**Example 2-14** We optimized, using mathematical models to improve the effectiveness and cost-effectiveness of vaccination programs, the combinations of timing, number, and efficacy of inoculations both for the vaccinated individual and for the overall population.

**Revised Example 2-14** **Using mathematical models to improve the effectiveness and cost-effectiveness of vaccination programs,** we optimized the combinations of timing, number, and efficacy of inoculations both for the vaccinated individual and for the overall population.

## 2.3 SENTENCE STRUCTURE AND TECHNICAL STYLE

### ➤ Use the first person

A generation and more ago, it was fashionable to avoid using "I" or "we" in scientific research papers because these terms were considered subjective, whereas the aim in science is to be objective. Contemporary scientific English, however, recognizes that science is not purely objective. It aims to engage the reader more and to make the writing interesting and active. Thus, in modern scientific English, use of the first person ("I" for one author; "we" for multiple authors) is actually preferred for describing what you did. If you look carefully at the most prestigious scientific journals, such as *Science* and *Nature*, you will find that most articles are written in first person and active voice.

| | | |
|---|---|---|
| 👎 | **Example 2-15** | It was hypothesized that spiders can regrow lost legs. |
| 👍 | **Revised Example 2-15** | **We hypothesized** that spiders can regrow lost legs. |
| 👎 | **Example 2-16** | The authors thank Peter Fefergon. |
| 👍 | **Revised Example 2-16** | We thank Peter Fefergon. |

There are a few exceptions to this rule in scientific documents. For example, use of the first person is more controversial in the Methods section than in the rest of a paper as in this section, materials and methods are usually the topic. In addition, often it may not have been the author(s) who performed a certain experiment but rather a technician or hired helper. Therefore, in the Materials and Methods section, use of third person and passive voice (the form of the verb used when the subject is being acted on rather than doing something) is often preferred.

### ➤ Use the active voice

Similar to preferring first person over third person in contemporary scientific documents, scientific authors now prefer the use of active voice over that of passive. When you write in passive voice, the subject is being acted on rather than actively doing something (for example: *Active Voice*: The dog bit the cat. *Passive Voice*: The cat was bitten by the dog.). If the passive voice is used excessively, writing becomes very dull and dense. Therefore, use the active voice rather than the passive voice when possible.

| | | |
|---|---|---|
| 👎 | **Example 2-17a** | Parrots were attacked by hawk B3. |
| 👍 | **Revised Example 2-17a** | Hawk B3 attacked parrots. |
| 👎 | **Example 2-17b** | Samples collected in June and July were analyzed by team A, No change in salt concentration was noted by the team. |
| 👍 | **Revised Example 2-17b** | Team A **analyzed** samples collected in June and July. The team **noted** no change in salt concentration. |

However, do not remove the passive voice completely. Use it when readers do not need to know who performed the action or when it sounds more natural, such as in the Materials and Methods section.

**Example 2-17c** After incubation, plasmids **were isolated** as described previously (3).

### ➤ Use the past tense for observations and specific interpretations

Many scientific authors seem to be confused about when to use past tense and present tense. Generally, you should use the past tense for observations, unpublished results, and specific interpretations (i.e., in the Results section of a lab report or scientific research paper).

**Example 2-18** The IV **caused** local irritation in 53% of the patients.

### ➤ Use the present tense for general rules and established knowledge

Use the present tense for general rules, accepted facts, and established knowledge.

**Example 2-19** Most deciduous trees had leaves.

**Revised Example 2-19** Most deciduous trees **have** leaves.

A sentence can also have mixed tenses, as is apparent in the next example.

**Example 2-20** Brown **reported** that the bacteria *Brucella* **cause** abortion in livestock.

Here, the experiment has been completed. "Reported" is therefore written in past tense. "Cause," however, is present tense because this part of the sentence is still true and considered established knowledge because it has been published.

### ➤ Write short sentences and aim for one main idea in a sentence

Short sentences are easier to understand than long sentences. Aim for sentences with an average number of words of about twenty to twenty-two. Make sure your sentences do not contain more than one main idea. Short, simple sentences tend to emphasize the idea contained in them. However, using only short sentences does not result in strong writing. Some sentences will be long to communicate complex ideas.

**Example 2-21** *(76 words)* When central venous IV lines were removed, skin samples from patients with IV line–related bloodstream infections were collected and, in 80% of these samples, bacteria with high DNA identity to those found in the bloodstream and IV lines were identified, whereas in 20% of the patients isolated bacteria had no or low DNA identity, indicating that most bloodstream infections in patients with central venous IV lines arise from contamination of IV lines or needles during insertion.

**Revised Example 2-21** When central venous IV lines were removed, skin samples from patients with IV line–related bloodstream infections were collected. In 80% of these samples, bacteria with high DNA identity to those found in the bloodstream and IV lines were identified. However, in 20% of the patients isolated bacteria had no or low DNA identity. These observations indicate that most bloodstream infections in patients with central venous IV lines arise from contamination of IV lines or needles during insertion.

### ➤ Use active verbs

Verbs are perhaps the most important part of an English sentence. Verbs make sentences direct and easy to follow. If you hide verbs by using nominalizations, that is, abstract nouns derived from verbs and adjectives, you will make your writing heavy and much more difficult for readers to understand. Use active verbs instead.

**Example 2-22** An increase in temperature occurred.

**Revised Example 2-22** Temperature **increased**.

In the revised sentence, the verb is active and strong. Thus, this sentence is simpler, more direct, and more efficient than when the action is nominalized.

### ➤ Avoid strings of nouns

Strings, or clusters of nouns that are strung together—often with adjectives—to form one term, often appear in scientific texts. However, such strings of nouns are awkward and often incomprehensible. When such terms appear one right after the other, it can be difficult to tell how they relate to each other and what the real meaning of the cluster is. Instead, use prepositions to link the nouns and adjectives. Prepositions add clarity to a phrase—they show more fully how the nouns relate to one another—and the meaning of your words becomes clearer.

**Example 2-23** porcine tracheal fluid samples

**Revised Example 2-23** fluid samples **from** the tracheae **of** pigs

**Example 2-24** The strips were exposed to Leishmaniasis disease patients' sera.

**Revised Example 2-24** The strips were exposed to **sera** *of* **patients** *with* **Leishmaniasis disease.**

Note that certain noun pairs and clusters, such as "water bath" and "sucrose density gradient," are accepted terms. Do not break such terms apart.

## ➤ Use clear pronouns and prepositions

### Pronouns

Unclear pronouns are one of the most common problems in scientific writing. If the pronoun that refers to a noun is unclear, the reader may have trouble understanding the sentence. Pronouns you use have to refer clearly to a noun in the current or previous sentence or paragraph. If the pronoun can refer to too many possible nouns, repeat the reference noun after the pronoun.

**Example 2-25** *Anaerobic organisms* typically live in the *intestines*. Thus, they are of interest to us.

**Revised Example 2-25a** Anaerobic organisms typically live in the **intestines**. Thus, **intestines** are of interest to us.

**Revised Example 2-25b** **Intestines** are of interest to us because *they* typically contain anaerobic organisms.

Sometimes the noun that a pronoun refers to has been implied but not stated. To clarify the reference, explicitly state the implied noun after the pronoun, as in the next example:

**Example 2-26** If a specimen is frozen in a bath containing dry ice and acetone, the water of the cell can be removed by sublimation and damage to the cell prevented. *This* is commonly used for preservation of cultures.

**Revised Example 2-26** If a specimen is frozen in a bath containing dry ice and acetone, the water of the cell can be removed by sublimation and damage to the cell prevented. *This technique* is commonly used for preservation of cultures.

## Prepositions

Most verbs can be used with more than one preposition. Be sure to choose the preposition that reflects your intended meaning.

**Example 2–27** **a** The human brain is sometimes **compared to** a computer.

**b** When we **compared** our results **with** those of Paulings et al. . . .

The most commonly (mis)used prepositions in scientific writing include:

| | |
|---|---|
| in connection **with** | compared **to/with** |
| in contrast **to** | correlated **with** |
| similar **to** | analogous **to** |
| comparison **of** A **with** B | comparison **between** A **and** B |

## ➤ Use correct parallel form

Lists and ideas that are joined by "and," "or," or "but" are of equal importance in a sentence and should be treated equally by writing them in parallel form. To write ideas in parallel form, the same grammatical structures are used.

**Example 2-28**

| | *subject* | *verb* | *prepositional phrase* |
|---|---|---|---|
| | **The metabolic rate of A** | **increased** | **2-fold** |
| but the | **metabolic rate of B** | **decreased** | **10-fold.** |

**Example 2-29**

| | *preposition* | *object* |
|---|---|---|
| The moss *P. patens* grows | **in** | **the mountains** |
| and | **at** | **sea level.** |

**Example 2-30** Prolonged fever, together with subcutaneous nodules in a child, could be due to an infection with a $Gram^+$ organism, but it could also be that the child suffers from rheumatic disease.

In this sentence, two equal ideas are connected by "but." However, these parallel ideas are not instantly apparent because the grammatical form of the first part of the sentence is not parallel to that of the second part. The second half of the sentence should be written in parallel form because it is equal in logic and importance.

| | | |
|---|---|---|
| 👍 | **Revised Example 2-30a** | Prolonged fever, together with subcutaneous nodules in a child, **could be due to** an infection with a Gram$^+$ organism, but it **could also be due to** rheumatic disease. |

This sentence can be further simplified:

| | | |
|---|---|---|
| 👍 | **Revised Example 2-30b** | Prolonged fever, together with subcutaneous nodules in a child, could be **due to** an infection with a Gram$^+$ organism or **due to** rheumatic disease. |

### ➤ Avoid faulty comparisons

Faulty comparisons and incomplete comparisons are two of the most common problems in scientific writing.

*Faulty comparisons* can arise due to ambiguous comparisons that "compare apples and oranges" instead of like items.

| | | |
|---|---|---|
| 👎 | **Example 2-31** | These results are similar to previous studies. |
| 👍 | **Revised Example 2-31a** | These **results** are similar to **the results of** previous studies. |
| 👍 | **Revised Example 2-31b** | These **results** are similar to **those of** previous studies. |

*Incomplete comparisons* may confuse readers because their intended meaning is unclear. Complete comparisons must include both the item being compared and the item it is being compared with.

| | | |
|---|---|---|
| 👎 | **Example 2-32** | RNA isolation is more difficult. |
| 👍 | **Revised Example 2-32** | RNA isolation is **more difficult than DNA isolation**. |

The overuse of "compared to" is equally confusing for readers. Use "than," not "compared to," for comparative terms such as "smaller," "higher," or "more."

| | | |
|---|---|---|
| 👎 | **Example 2-33** | We found more fertilized eggs in buffer A compared to buffer B. |
| 👍 | **Revised Example 2-33** | We found **more** fertilized eggs in buffer A **than in** buffer B. |

### ➤ Avoid errors in spelling, grammar, and punctuation

Common errors in writing include (1) spelling, (2) punctuation, (3) word omissions, (4) subjects and verbs that do not make sense together, (5) subjects and verbs that do not agree, and (6) unclear modifiers. When one of these errors appears, the reader is slowed down and may even need to re-read the sentence to figure out the intended meaning. By themselves, occasional spelling, grammar, or punctuation errors will not bother readers nearly as much as errors in the larger structures, such as logical gaps and misplaced information. If the smaller errors are frequent, however, they can easily add up to major annoyances and discredit your work. These errors can be avoided by carefully proofreading and double-checking the manuscript. Note that the spell check function in your word processing program has only limited applicability because many words such as "to," "two," and "too" will not be caught when they are misspelled or chosen incorrectly.

Many scientific authors are confused about such words as "data," "spectra," and "media." "Data," "spectra," and "media" are the plural forms of "datum," "spectrum," and "medium" and thus should be treated as plural nouns. (Note that some dictionaries do accept use of both singular verbs and plural verbs with these words.)

---

**Example 2-34** Our data suggests that Klein's hypothesis is correct.

---

**Revised Example 2-34** Our **data suggest** that Klein's hypothesis is correct.

---

Another common mistake is dangling or misplaced modifiers.

---

**Example 2-35** Having tested positive for HIV, we disqualified the patients for participation in the study.

---

This modifier is unclear because it modifies "we." It sounds as if "we" tested positive for HIV.

---

**Revised Example 2-35a** **Having tested positive for HIV, the patients** were disqualified for participation in the study.

---

**Revised Example 2-35b** **Patients that tested positive for HIV** were disqualified for participation in the study.

---

## SUMMARY

### INDIVIDUAL WORDS

1. Use precise words.
2. Use simple words.
3. Omit unnecessary words and phrases.
4. Avoid too many abbreviations.
5. Use correct terminology and nomenclature.

### WORD LOCATION

1. Establish importance.
2. Place old, familiar, and short information at the beginning of a sentence.
3. Place new, complex, or long information at the end of a sentence.
4. Get to the subject of the main sentence quickly.
5. Place the verb immediately after the subject and the object right after the verb.

### SENTENCE STRUCTURE AND TECHNICAL STYLE

1. Use the first person.
2. Use the active voice.
3. Use the past tense for observations and specific interpretations.
4. Use the present tense for general rules and established knowledge.
5. Write short sentences and aim for one main idea in a sentence.
6. Use active verbs.
7. Avoid strings of nouns.
8. Use clear pronouns.
9. Use correct parallel form.
10. Avoid faulty comparisons.
11. Avoid errors in spelling, punctuation, and grammar.

## PROBLEMS

### PROBLEM 2-1 Precise Words

**Find the nonspecific terms in the following sentences. Replace the nonspecific choices with more precise terms or phrases. Note that it is not necessary to change the sentence structure; just replace the individual words. Guess or invent something if you have to.**

1. Plants were kept in the cold overnight.
2. Some exoplanets orbit multiple stars.
3. The population of bivalve molluscs per square meter decreased markedly between 1999 and 2001.

**PROBLEM 2-2 Simple Words**

**Improve the word choice in the following examples by replacing the underlined terms or phrases with simpler word choices. Do not change the sentence structure; just change the words.**

1. These data substantiate our hypothesis.
2. We utilized UV light to induce mutations in the plants.
3. The differences in our results compared to those of Reuter et al. (1995) can be accounted for by the fact that different conditions were used.
4. For the purpose of examining cell migration, we dissected mouse brains.
5. An example of this is the fact that population densities differ substantially.

**PROBLEM 2-3 Redundancies**

**Improve the word choice of the underlined words in the following examples by removing any redundancies and unnecessary words and phrases. Do not change the sentence structure.**

1. The doubling rate appeared to be quite short.
2. It is known that homologous recombination is the preferred mechanism of DNA repair in yeast.
3. Often, jewel weed can be found to grow in close proximity to poison ivy. (Two corrections needed.)
4. Upon heat activation, filament size increased, and the number of buds decreased. Both the increase in filament length and the decrease in the number of buds were only seen for cytokinin mutants.

**PROBLEM 2-4 Word Placement and Flow**

**Rewrite the following paragraph. Place words such that the reader can easily follow the logic flow of the message.**

Rainwater often picks up carbon dioxide, resulting in a weak solution of carbonic acid. A cave is formed when such rainwater trickles into the ground in areas with a high limestone content. Carbonic acid slowly dissolves the limestone. As more and more limestone dissolves, the cave grows underground. When a cave's ceiling gets eroded and collapses, a sinkhole forms.

**PROBLEM 2-5 Word Placement and Flow**

**Write a paragraph using the list of facts provided. To create good flow, place words carefully at the beginning and end positions of sentences.**

- fleas transmit plague *bacillus* to humans
- *bacilli* migrate from bite site to lymph nodes
- name "bubonic plague" arises because buboes = enlarged nodes

**PROBLEM 2-6 Word Placement and Flow**

1. **Construct a paragraph about thermophiles using the list of facts provided. Pay attention to good flow of the message by considering word placement.**
2. **What does the reader expect to see next after having read the last sentence of your paragraph?**

*Thermophiles*

- microorganisms
- temperature range for growth between 45°C and 70°C
- found in hot sulfur springs
- cannot grow at body temperature
- not involved in infectious diseases of humans
- mechanism to resist elevated temperature unclear

**PROBLEM 2-7 Active Verbs**

**Put the action in the verbs in the following sentences.**

1. An increase in transplant rejection occurred.
2. Two measurements of amyloid plaques were obtained for each brain.
3. Our results showed protection of the dogs by the vaccine.
4. To determine whether cell migration is occurring, we dissected mouse brains.
5. Buffalo, elephant, and black rhino abundance all show a rapid decline after 1977.

**PROBLEM 2-8 Pronouns**

**The following sentences contain unclear pronouns. Improve the clarity of these sentences by repeating one or more words from the previous sentence, rearranging the sentence, or adding a category or implied term instead of the underlined pronoun.**

1. A few microorganisms such as *Mycobacterium tuberculosis* are resistant to phagocytic digestion. <u>This</u> is one reason why tuberculosis is difficult to cure.
2. Our findings indicate that binding decreases about 10-fold when temperature increases from 15°C to 25°C. <u>This</u> suggests that binding among the particles is not due to ionic interactions.
3. The color was achieved using new methods and concepts developed in our laboratory to distinguish specific cell wall components. <u>This</u> in turn was only possible through heat induction.

**PROBLEM 2-9 Parallelism and Comparisons**

**Correct the faulty parallelism or comparisons in the following sentences.**

1. The pathogenesis observed in other cells, such as circulating monocytes, may differ from endothelial cells.
2. Diabetes can be affected both by exercise and diet.
3. We observed a peak for mutant A that was higher than the other mutants.
4. Compared to the other mammals, the male dolphin was larger.
5. Many wasps can sting more often compared to honey bees.

CHAPTER 3

# Fundamentals of Scientific Writing—Composition

THE OBJECTIVE OF THIS CHAPTER IS TO CONVEY:

- How to construct an effective paragraph
    - How to organize a paragraph
    - How to make a paragraph coherent
- How to create good flow using word location, key terms, and transitions
- How to emphasize important ideas and signal subtopics
- How to condense text

More important than using good style in scientific writing is to order and structure your sentences, paragraphs, and sections logically. This chapter adds to the basic scientific writing rules discussed in Chapter 2 by addressing the larger structures (paragraphs) of documents.

## 3.1 PARAGRAPH FLOW

The basic unit of composition in the English language is the paragraph. For readers to follow the flow and understand the story of a paragraph, it has to be composed unambiguously and logically. The following example presents a paragraph in which the author has not paid much attention to the needs of readers, be they peers or an educated lay audience:

**Example 3-1** **(1)** Volcanic ash adsorption poses a great environmental hazard. **(2)** The deposition of this ash and the subsequent draining of its volatiles is a rapid route by which elements and ions are delivered to the ground (3–5). **(3)** Due to similar magma types, leachate content from volcanoes in close proximity to each other appears alike. **(4)** The greatest hazard to the environment is posed by magmas with relatively high halogen content, and many hazardous leachate fluoride concentrations are found in volcanoes with high $F/SO_4^{2-}$ ratios. **(5)** Particles <2 mm across seem to have a greater adsorption, and therefore a high leachate hazard exists even with low ashfall (7, 8). **(6)** Under high humidity conditions, gas accumulation can be further increased as bigger sulphuric acid droplets make contact with ash particles more likely (12). **(7)** The measuring and reporting of leachate results should be standardized.

*(Reference: Witham, 2005)*

If you catch yourself reading this paragraph more than once, you are not alone. You may think that you have not paid enough attention and start reading it over. Some readers may even consider themselves not smart enough to understand the topic. The individual sentences are intelligently composed and free of grammatical errors. The sentences are also not overly long or complex. The vocabulary is professional but not beyond the scope of the educated general reader. Nonetheless, most of you arrive at the end of the paragraph without fully understanding what the author is saying. The problem lies not with you but with the author. The paragraph has not been structured or organized. Sentences seem to be put together at random and not in any logical order. Important information, such as transitions and logical connections, seem to have been left out. Consequently, the paragraph does not flow. In short, the author has not composed the paragraph with the reader in mind.

Let us look at the revised version of Example 3-1a:

---

**Revised Example 3-1a**

**(1)** Volcanic ash adsorption poses a great environmental hazard. **(2)** The deposition of this ash and the subsequent leaching of its volatiles is a rapid route by which elements and ions are delivered to the ground (3–5). **(3′)** *Adsorption can be influenced by magma type, particle size, and humidity conditions (7).* **(3)** *For example*, leachate content from volcanoes in close proximity to each other appears alike due to similar magma types. **(4)** *In fact*, the greatest hazard to the environment is posed by magmas with relatively high halogen content, and many hazardous leachate fluoride concentrations are found in volcanoes with high $F/SO_4^{2-}$ ratios. **(5)** *Aside from magma type*, particles <2 mm across seem to have a greater adsorption, and therefore a high leachate hazard exists even with low *ash deposition* (7, 8). **(6)** *In addition*, under high humidity conditions, *adsorption* can be further increased as bigger sulphuric acid droplets make ash adsorption more likely (12). **(7)** *Ideally*, the measuring and reporting of leachate results should be standardized.

*(Reference: Witham, 2005)*

---

After looking at this revision, you may now recognize the lack of structure and organization in the original paragraph. You can see that an important missing link was sentence 3′. It ties sentences 3 through 6 together and links them to the beginning of the paragraph. Adding the transitions "for example," "in fact," "aside from," "in addition," and "ideally" logically connects the ideas in the paragraph. These transitions are missing in the original paragraph. Replacing "ashfall" with "ash deposition" and "gas accumulation" with "adsorption" helps keep the reader focused on the topic because terms are used more consistently throughout the paragraph.

Although the revision greatly improved the paragraph, we could revise it even more:

---

**Revised Example 3-1b**

**(1′)** Volcanic ash adsorption poses a great environmental hazard *because adsorbed volatiles can be rapidly deposited and subsequently leached into the ground.* **(2′)** *Adsorption can be influenced by magma type, particle size, and humidity conditions (7).* **(3)** *For example*, leachate content from volcanoes in close proximity to each other appears alike. **(4)** *In fact*, the greatest hazard to the environment is posed by magmas with a relatively high halogen content and by magmas with high $F/SO_4^{2-}$ ratios in which many hazardous leachate fluoride concentrations are found. **(5)** *Aside from* *magma type*, particles <2 mm across seem to have a greater adsorption, and therefore a high leachate hazard exists even with *low ash deposition* (7, 8). **(6)** *In addition*, under high humidity conditions, *adsorption* can be further increased as bigger sulphuric acid droplets make ash adsorption more likely (12). **(7′)** *Unfortunately*, *currently no uniform leachate analysis methods exist and thus data is difficult to compare.* **(8)** *Ideally*, the measuring and reporting of leachate results should be standardized.

*(Reference: Witham, 2005)*

---

The flow of the paragraph has been further improved in Revised Example 3-1b. Sentences 1 and 2 have been combined into sentence 1′, making their relationship clearer through the use of "because." Another link, sentence 7′, has been placed before the last sentence to more logically connect it to the rest of the paragraph.

To make a paragraph clear, it needs to be written such that readers can understand the logic and story at once, regardless of the scientific content. A well-constructed paragraph must not only be organized, it must also be coherent. In addition, important ideas should be emphasized, and subtopics should be signaled.

## 3.2 PARAGRAPH CONSTRUCTION AND ORGANIZATION

### ➤ Organize your paragraphs

In scientific writing, as in writing ordinary prose, paragraphs are typically constructed of three principal parts: a topic sentence, which introduces the overall idea of the paragraph; supporting sentences, which provide the details on the topic; and a concluding sentence, which summarizes the information presented or sets the stage for the next topic or paragraph. (Note that short paragraphs may not contain a concluding sentence, although long paragraphs usually do. In addition, in less formal writing, or when providing technical instructions, paragraphs do not always follow this formal construction.)

Sentences within a formal paragraph need to be logically organized and positioned. Each of these paragraphs has two important power positions: the first sentence and the last sentence. Accordingly, important information should be placed in these positions rather than in the remaining sentences.

### ➤ Use a topic sentence to provide an overview of the paragraph

A well-written paragraph generally gives an overview first and then goes into detail. The topic sentence states the central topic or message of the paragraph and guides the reader into the paragraph. The end or stress position of the topic sentence highlights the topics that the author wants readers to follow in the rest of the paragraph. A topic sentence may also contain a transition from the previous paragraph or section. The details found in the remaining, supporting sentences are organized logically and consistently to explain the message provided by the topic sentence.

**Example 3-2** *Volatile organic compounds (VOCs)* are released from many **manmade** and **natural sources**. **Manmade sources** range from cars and industrial sources to construction materials, heaters, and other consumer products. **Natural sources** responsible for biogenic VOC emissions include mainly trees as well as fungi and microorganisms. The exact contribution of **manmade** versus **natural sources** of VOCs has not been determined, but clearly depends on locale and environment.

The topic of the above paragraph is "volatile organic compounds (VOCs)." The pattern of organization—that is, the order of the remaining sentences—is not random but follows the order the items are listed: *manmade* first, then *natural*.

### ➤ Use consistent order

To keep paragraphs organized authors also need to pay attention to keeping a consistent order of topics. If you list items in a topic sentence and then describe them in the remaining sentences at the paragraph level, keep the same order.

**Example 3-3** In response to a foreign macromolecule, five different immunoglobulins can be synthesized: **IgA, IgD, IgE, IgG,** or **IgM**. **IgA** is the major class in external secretions, such as saliva, tears, and mucus. Thus, **IgA** serves as a first line of defense against bacterial and viral antigens. **IgD** is an antigen receptor on B cells not yet exposed to antigens. **IgE** protects against parasites. **IgG** functions as the main immunoglobulin in serum. **IgM** is the first class to appear following exposure to an antigen.

### ➤ Use consistent point of view

Be consistent in your point of view or person. Switching from one style to another within a paragraph or document disorients the reader. To create a consistent point of view, and thus good flow of a paragraph or section, use the same subject or person throughout your paragraphs and sections.

**Example 3-4**

**(1)** These findings suggest that <u>people</u> returning from tropical countries who show very itchy linear or serpiginous tracked skin eruptions should be tested for larvae of animal hookworms. **(2)** However, <u>you</u> may also suffer from a specific type of rash.

In sentences 1 and 2, the point of view is not consistent. For the second subject, the person is switched to *you*. Switches like this are very disruptive to the paragraph and disorienting for the reader.

**Revised Example 3-4**

**(1)** These findings suggest that <u>people</u> returning from tropical countries who show very itchy linear or serpiginous tracked skin eruptions should be tested for larvae of animal hookworms. **(2)** However, **people** may also suffer from a specific type of rash.

## ➤ Make your sentences cohesive

Within a paragraph, the sentences need to be cohesive so their logical layout fits neatly together. When authors arrange sentences to be cohesive, they consider word location. Good word location creates good "flow" of a paragraph.

**Example 3-5**

Important pathogens can be found in the genus ***Yersinia***. ***Yersinia*** contains several **species. One species,** *Y. pestis*, is the cause of bubonic **plague**. The **plague** bacillus infects lymph nodes near the site of infection to produce buboes.

The reader conceives sentences as cohesive if the information provided at the beginning of a sentence can be found toward the end of the respective previous sentences (see Basic Rules in Chapter 2, Section 2.2).

Placing information provided at the end of the sentences toward the beginning of the next sentence is not the only way to provide paragraph cohesion. Cohesion can also be created by providing a consistent point of view.

**Example 3-6**

**Rhubarb** is a frequently used Chinese herbal medicine. **It** is used to treat various ailments including constipation, inflammation, and cancer.[1,2] As a drug, **rhubarb** is made up of the roots and rhizomes of three members of the *Polygonaceae* family: *Rheum officinale, R. palmatum*, and *R. tanguticum*. Different **rhubarb** species show substantial differences in purgative effects and chemical compositions. However, **they** are similar in physical appearance and thus difficult to distinguish.

Here, information in the topic position of each sentence refers back to the topic position of the topic sentence.

Many paragraphs contain a mixture of word location. For some sentences, information in the topic position may refer back to the end position of the previous sentence. For other sentences, information may be written from the point of view of the old information and refer back to the topic position of the topic sentence, as in the next example.

**Example 3-7** There are positive and negative factors that should be

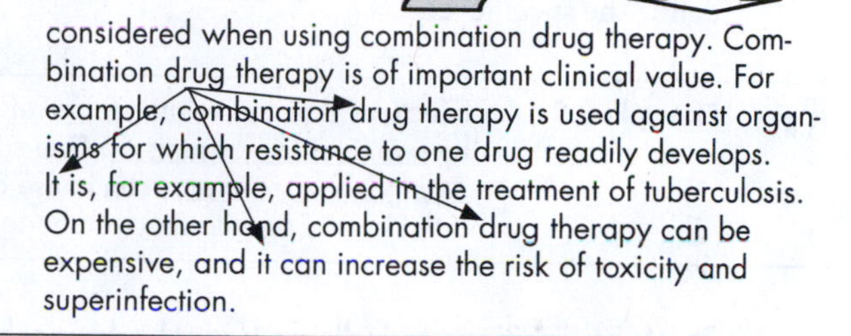

considered when using combination drug therapy. Combination drug therapy is of important clinical value. For example, combination drug therapy is used against organisms for which resistance to one drug readily develops. It is, for example, applied in the treatment of tuberculosis. On the other hand, combination drug therapy can be expensive, and it can increase the risk of toxicity and superinfection.

Constructing paragraphs in which mixed word locations occur is fine—as long as the links do not get too muddled or tangled to risk losing the reader.

### ➤ Use key terms to create continuity

Readers consider a paragraph to be coherent if they can quickly find the topic of each sentence and if they see how the topics are a related set of ideas. Thus, a coherent paragraph needs to consist of a series of sentences that lead logically from one to the next. Cohesion is achieved not only through the use of word location but also through key terms and transitions.

Key terms are words or short phrases used to identify important ideas in a sentence, a paragraph, and the paper as a whole. Usually, key terms are used to identify your main points in the topic sentence. Repeating and linking terms ensures that the topic of the work cannot be missed and that relationships between topics are clear.

To avoid confusing readers, key terms should not be changed. If you deliberately repeat key terms, your main points are emphasized and you create continuity.

**Example 3-8** Worms are a great system for studying how diabetes, age, and obesity are connected because these organisms have a well-conserved insulin signaling pathway that affects all of these factors. The entire genome of *C. elegans* has also been sequenced.

Readers unfamiliar with this particular topic may not know that "worms" and "*C. elegans*" refer to the same thing. They may be confused when two different terms are used.

**Revised Example 3-8**

***Caenorhabditis elegans*** are a great system for studying how diabetes, age, and obesity are connected because these organisms have a well-conserved insulin signaling pathway that affects all of these factors. The entire genome of ***C. elegans*** has also been sequenced.

When you need to shift from a category term to a more specific term or the other way around, key terms should be linked so continuity is not lost and the paragraph stays coherent. To link key terms, use the category term to define the specific term.

**Example 3-9**

Infectious diseases related to travel may be caused by gram-positive organisms. **One such organism,** *Staphylococcus aureus*, can cause cellulitis, purulent arthritis, and suppurative lymphadenitis.

### ➤ Use transitions to indicate logical relationships between sentences

When word location and key terms alone are not sufficient to link sentences within a paragraph or section, transitions need to be used. Transitions are words, phrases, or even sentences that logically relate sentences to each other and to the overall story, such as *in addition*, *thus*, and *It is because of X*... (see Table 3.1 for more examples.) Transitions should be placed at the beginning of a sentence for strongest continuity and are usually set off by a comma.

**Example 3-10**

We determined sensitivity to external stimulation in hibernating ground squirrels. We measured active, torpid, and interbout arousal states daily by measuring core body temperature.

In this example, the logical relationship between sentences 1 and 2 is not clear because a transition is missing. Once the transition is added in, the relationship between the two sentences becomes obvious.

**Revised Example 3-10**

We determined sensitivity to external stimulation in hibernating ground squirrels. **For this purpose**, we measured active, torpid, and interbout arousal states daily by measuring core body temperature.

**TABLE 3.1 Transition Words, Phrases, and Sentences**

| USED FOR | TRANSITION WORDS | TRANSITION PHRASE | TRANSITION SENTENCE |
|---|---|---|---|
| Addition | again, also, further, furthermore, in addition, moreover | In addition to X, we . . .<br>Besides X . . . | Further experiments showed that . . . |
| Comparison | also, in the same way, likewise, similarly, etc. | As seen in . . . | When A is compared with B . . .<br>As was reported by . . . |
| Concession | clearly, evidently, obviously, undeniably | | Granted that X is . . . |
| Contrast | but, however, nevertheless, nonetheless, still, yet | In contrast to A . . .<br>X on the other hand . .<br>Despite X . . . ,<br>Unlike X. . . | One difference is that . . .<br>Although X differed . . . |
| Example | for example, specifically | To illustrate X . . . | An example of this is that . . . ,<br>That is . . . |
| Explanation | Here | Because of X . . .<br>In this experiment . . . | One reason is that . . .<br>Because X is . . . |
| Purpose | for this purpose, to this end | For the purpose of . . . | The purpose of X was to . . .<br>To determine XYZ, we . . . |
| Result | consequently, generally, hence, therefore, thus | As a result of . . . | Evidence for XYZ was that . . .<br>Analysis of ABC showed that . . . |
| Sequence/time | after, finally, first, later, last, meanwhile, next, now, second, then, while | After careful analysis of . . .<br>During centrifugation, . . . | After X was completed, . . .<br>When we . . . |
| Summary | in brief, in conclusion, in fact, in short, in summary | To summarize (our results), . . . | |

**Strength of transition**

## 3.3 CONDENSING

### ➤ Make your writing concise

A well-written paragraph should be concise. Condensing a passage often needs to be done in combination with other techniques:

1. Emphasize important information.
2. Deemphasize or omit less important information.
3. Replace or omit words and phrases.

### Establishing Importance

The key in condensing is to distinguish between important and less important information. Often, authors are too close to their own writing to be able to do so. Instead, they consider everything important and have a hard time prioritizing statements. To help you prioritize, it is often critical to have someone else read your draft. The best critics in this respect are your peers, instructors, or trained scientists. These people are able to guide you in putting levels of importance into better perspective.

Once you have decided on these levels of importance, distinguish information by highlighting important points, deemphasizing less essential ones, and omitting nonessential phrases. Important information can be highlighted by placing it into a power position and/or by signaling it, such as "Most importantly, . . ." or "The key finding of this study was . . ." Less important information can be deemphasized by omitting or subordinating it—for example, by placing it in a subordinate clause.

**Example 3-11** ***Although*** *the migration and expansion of the human population coincides with the extinction of large animals,* it is unclear if environmental changes were also associated with the onset of the last glacial cycle.

### Words and Phrases that Can Be Omitted or Condensed if Needed

1. Omit overview words and phrases such as:

| | |
|---|---|
| described | noted |
| noticed | observed |
| reported | seen |

**Example 3-12** Jones et al. (1996) showed that intracellular calcium is released when adipocytes are stimulated with insulin.

*(16 words)*

**Revised Example 3-12** Intracellular calcium is released when adipocytes are stimulated with insulin (Jones et al., 1996).

*(10 words)*

2. Omit phrases and sentences that tell your reader what a sentence/paragraph is about.

**Example 3-13** Products were verified by gel electrophoresis. The results are presented in Figure 1.

*(13 words)*

**Revised Example 3-13** Products were verified by gel electrophoresis **(Fig. 1)**.

*(6 words)*

3. Omit "It . . . that" phrases. Most of these phrases are pointless fillers and can be omitted entirely. If the idea in the phrase is essential, replace the phrase with a shorter version.

**Examples of "It . . . that" phrases:**

| | |
|---|---|
| It is interesting to note that . . . | **omit** |
| In light of the fact that . . . | **replace (because)** |
| It is possible that . . . | **reword (. . . may . . .)** |
| It has been reported that . . . | **omit or replace (Taylor reported that . . ., or [reference])** |

4. Change negative to positive expressions. This change usually results in shorter sentences. Moreover, readers prefer to read positive things, not negative ones. Above all, avoid double negatives, which can easily confuse readers.

**Examples of changing from negative to positive:**

| **negative** | **positive** |
|---|---|
| do not overlook | note |
| not different | similar |
| not many | few |
| not the same | different |

5. Omit excessive detail. Detail that can be inferred or is unimportant should be omitted.

| | | |
|---|---|---|
| | **Example 3-14** | Using a 1-ml tip, 750 μl of the aliquot were removed into a new Eppendorf tube after incubation. |
| | **Revised Example 3-14** | After incubation, 750 μl of the aliquot were removed. |

6. Do not overuse hedges or intensifiers. Hedges are cautious adjectives, adverbs, or verbs such as:

   usually, often, possibly, perhaps, some, most, many,
   appear, could, indicate, may, seem, suggest

   Intensifiers are adjectives, adverbs, or verbs that are used to strengthen nouns or verbs, such as:

   always, clearly, certainly, basic, central, crucial, show, prove

| | | |
|---|---|---|
| | **Example 3-15** | Figure 5 clearly shows that the protein was absent in the fraction. |
| | **Revised Example 3-15** | Figure 5 shows that the protein was absent in the fraction. |

Within lab reports, scientific articles, and posters, intensifiers hedges are usually avoided because data should be presented objectively. However, in proposals and job applications, intensifiers are common because these documents try to persuade rather than present.

**Example 3-16** ***Within a CV/résumé:***

- Outstanding computer skills
- Pivotal role in ABC study
- Proven leadership ability

## SUMMARY

### BASIC RULES—COMPOSITION

1. Organize your paragraphs.
2. Use a topic sentence to provide an overview of the paragraph.
3. Use consistent order.
4. Use consistent point of view.
5. Make your sentences cohesive.
6. Use key terms to create continuity.
7. Use transitions to indicate logical relationships between sentences.
8. Make your writing concise.

## PROBLEMS

### PROBLEM 3-1 Paragraph Organization

**The following paragraph is about the migration of salmon. Although sentences 1 and 2 describe the two possible methods salmon may use to find their way, the paragraph does not have a topic sentence. Write a clear topic sentence for this paragraph. The topic sentence should state the message of the paragraph (the different methods salmon use to find their way during their homeward journey in the fall). In your topic sentence, make the topic the subject of the sentence. Also, make sentence 2 parallel to sentence 1.**

**(1)** Salmon use geomagnetic imprinting to return to their freshwater birthplace to spawn. **(2)** As described by Stabell et al. (15), olfactory cues also play a role in guiding adult salmon back to the stream of their birth. **(3)** It is unclear whether salmon also use cues other than geomagnetic and chemical imprinting to orient themselves.

### PROBLEM 3-2 Paragraph Construction

**Construct a paragraph on cancer cells using the list of facts in the order provided. In your writing, pay attention to writing a good topic sentence and to using good word location. Employ paragraph consistency, key terms, and transitions. Also consider other basic rules such as parallel form and correct pronouns and prepositions.**

*Cancer cells*

- Are malignant tumor cells
- Differ from normal cells in three ways:
    - They dedifferentiate—for example: ciliated cells in the bronchi lose their cilia
    - Metastasis is possible—travel to other parts of body. New tumor growth
    - Rapid division—do not stick to each other as firmly as do normal cells.

**PROBLEM 3-3 Constructing a Paragraph**

**Compose a short passage using the following bullet points about DEET (N,N-diethyl-meta-toluamide):**

- Most common active ingredient in insect repellents
- Developed by the U.S. military in the 1940s
- A few reports of adverse effects in humans (Osimitz and Grothaus, 1995; Sudakin and Trevathan, 2003)
- Easily absorbed into the skin
- Considered safe if used correctly
- Heavy and frequent dermal exposure can lead to skin irritation and, in rare cases, death, especially in young children (Osimitz and Grothaus, 1995; Sudakin and Trevathan, 2003)
- Neurological damage and death can result from ingestion (Osimitz and Grothaus, 1995; Osimitz and Murphy, 1997)

**PROBLEM 3-4 Constructing a Paragraph**

**Use the following bullet points to create a meaningful paragraph on polar ice reduction. In composing this cohesive paragraph, create good flow by considering word location, key terms, and transitions.**

- Global warming has been a major concern over the past decades
- Ocean levels have risen 15–20 cm in the past century
- By 2100, ocean levels are expected to rise another 13–94 cm (Haug, 1998)
- As our climate warms, ocean temperatures rise
- Water expands at higher temperatures, leading to sea level rise
- Polar ice melt increases temperature rise on Earth whereas ice reflects sunlight, oceans absorb light and heat
- As temperature rises, glaciers and ice sheets melt
- Temperature rise has been accelerated by humans

**PROBLEM 3-5 Condensing**

**Condense the following paragraph. Try to make your revised paragraph less than 30 words.**

Our results indicate that between 5°C and 25°C, the oxygen production of the moss *Brachythecium rutabulum* and that of *Physcomitrella patens* did not significantly change. *P. patens* also showed no significant change in oxygen production at higher temperatures. However, *B. rutabulum* clearly displayed a pronounced increase in oxygen production at temperatures above 25°C.

*(53 words)*

**PROBLEM 3-6 Paragraph Revision**

**Compose a paragraph on a scientific topic of your interest. Ensure that it flows well, check that no writing principles have been violated, and condense it as much as possible. Then, give it to a fellow student to edit and revise.**

# PART II

# Working with References and Data

CHAPTER 4

# Literature Sources

THIS CHAPTER PROVIDES DETAILS ABOUT:

- Locating, selecting, managing, and using references
- Formatting citations and reference lists
- How and where to cite the work of others—published in journals or on the Web—within scientific texts
- What constitutes plagiarism
- Keeping track of ideas and references
- How to paraphrase

Science builds on acquired and documented knowledge. Therefore, being able to work with references is important for two reasons: (1) to identify appropriate information of others and (2) to incorporate relevant information in your own writing. This chapter deals with both of these key aspects of scientific communication.

## 4.1 SEARCHING THE LITERATURE

Reading and understanding scientific literature, writing laboratory reports, composing a research paper, or preparing a review article or thesis typically requires you to be able to search for appropriate information, especially online. Using such information allows you to apply up-to-date research in your laboratory reports and to compose essays that would not normally be available in textbooks. In addition, if you submit coursework that includes references and information from relevant and recent publications, it shows that you have made an effort to research your work thoroughly and to validate it by relating it to contemporary research.

In the professional world, scientists need to be familiar with previously published findings and how to find them. They use this information to design new experiments, to cite sources of data that they use in order to give credit and provide evidence, and to show how their interpretations integrate with published scientific knowledge overall. Scientists not familiar with this process have little credibility among their peers. It is therefore essential to familiarize yourself with the use of the scientific literature.

### ➤ Know where to find sources

Several sources of scientific information exist in a variety of mediums. Because of the time it takes to publish a book, books usually contain more dated information than the most recent journals and newspapers. A huge amount of information is also available online, but it may not be immediately clear what resource to use for the purpose of composing a scientific document. Internet search engines may not prove ideal for obtaining information to compose a scientific document, because you may find it difficult to distinguish relevant, reliable, and authoritative information and to obtain full-length publications. Internet sites such as Google Scholar (http://scholar.google.com) provide access to primary literature, that is, peer-reviewed research articles. Generally, however, bibliographic databases such as PubMed as well as your university library offer a more reliable and structured way of finding good-quality information. Such databases can provide links to the full text of scholarly journals, in which you can find the most up-to-date information on research in industry and academia.

### ➤ Distinguish between primary, secondary, and tertiary sources

Work published in peer-reviewed, scholarly journals has undergone a rigorous evaluation process by experts in the field to maintain standards and provide credibility. Such scientific journal articles are considered primary sources because they typically report results for the first time—in contrast to secondary sources, which analyze and discuss the information provided by primary sources, and tertiary sources, such as textbooks and dictionaries, which compile and reorganize information provided in mainly secondary sources (see also Table 4.1). Scientific information available in scholarly journals can be found through your library or by searching online databases such as PubMed, HighWire, MEDLINE, and the Web of Science using key words, author, source, author's affiliation, cited author, cited work, and cited year.

**TABLE 4.1 Definitions and Examples of Primary, Secondary, and Tertiary Sources**

| SOURCE | DEFINITION | EXAMPLE |
|---|---|---|
| primary | original, peer-reviewed publication of a scientist's new data, results, and theories; report results for the first time | Scientific journal articles; theses, dissertations; conference proceedings; speeches |
| secondary | analyze and discuss the information provided by primary sources | Review articles; literary criticisms; some textbooks; commentaries |
| tertiary | compile and reorganize information provided in mainly secondary sources | Textbooks (some may also be secondary); dictionaries; manuals; Wikipedia |

### ➤ Become familiar with the most important science databases

The most important science databases are:

| | |
|---|---|
| **PubMed** | PubMed comprises more than 20 million citations for biomedical literature from MEDLINE, life science journals, and online books. Citations may include links to full-text published works from PubMed Central and publisher websites. (https://www.ncbi.nlm.nih.gov/pubmed) |
| **MEDLINE** | MEDLINE is produced by the US National Library of Medicine and is part of the National Institutes of Health. This database covers more than 4,000 journal titles and is international in scope. Broad coverage includes basic biomedical research and clinical sciences. (http://clarivate.libguides.com/webofscienceplatform/bci) |
| **Google Scholar** | This free site indexes the full text of most peer-reviewed online journals of Europe and America's largest scholarly publishers. (https://scholar.google.com) |
| **HighWire** | This archive is the largest one for free full-text science articles. It hosts more than 3,500 peer-reviewed journals as well as e-books, conference proceedings, and databases. (https://www.highwirepress.com) |
| **Web of Science** | The ISI Citation Databases collectively index more than 8,000 peer-reviewed journals. They provide Web access to Science Citation Index Expanded, which covers 6,300 international science and engineering journals. (https://login.webofknowledge.com) |

Other key science databases include:

| | |
|---|---|
| **Current Contents** | Current Contents Science Edition covers all the science editions of the Current Contents Search database in one package. (http://www.ovid.com/site/catalog/databases/862.jsp) |
| **BIOSIS** | Biosis, the online version of Biological Abstracts and Biological Abstracts, Reports, Reviews, and Meetings, contains literature references from all of the life sciences. This is the premier database for coverage of botany research. (http://clarivate.libguides.com/webofscienceplatform/bci) |
| **Scopus** | Scopus provides broad international coverage of the sciences and social sciences, indexing 14,000 journals. (https://www.elsevier.com/solutions/scopus) |

Some of these databases require subscription or registration, whereas others are free (e.g., Google Scholar).

In many cases, search results link directly to journal articles, and most users will only be able to access a brief summary of the articles. In these cases, you may have to request articles through interlibrary loans or directly from the publisher. Sometimes you may have to pay a fee to access entire articles. Other journals, such as *PLOS*, provide free access to complete articles online.

A longer list of major databases and search engines for finding and accessing articles in academic journals can be found at http://en.wikipedia.org/wiki/Academic_databases_and_search_engines. Note that many societies also have lists of useful journals for their disciplines (e.g., Society of Neuroscience, Society for Conservation Biology, American Society for Cell Biology).

## 4.2 SOURCE MATERIAL

### ➤ Use appropriate search terms

If you already know the details of a journal article (authors, title, date of publication, journal name, volume number, etc.), you can use the journal search to check if the library holds the article by searching for the title of the journal in which the article is published. To identify journal articles on a particular topic, you need to use a database. Searching for an article is easier if you already know a bit about the topic, for example, from reading the course textbook, lab manual, and lecture notes. This means you will already know some of the jargon and key terms for the topic before you start. In this case, you can search for articles using key words and refine your search by changing or combining key words, and by limiting databases or time of publication.

When a topic is mentioned in your textbook or lectures, you can search for a cited reference to find out more details about the topic. Such cited references in turn might lead you to other cited sources. When reading for your essay/report, you may also find one person's research mentioned frequently in the textbook, and you can then search for works and publications of this person as an author.

Many databases have an online thesaurus that you can consult to help in your search. If needed, ask a librarian for help in best use and combination of key words or search terms or to refine your search further. You can also use wildcards such as the "*" to find words containing the same root or if you are unsure about the spelling and conjunctions such as "and" and "or" to expand or refine your search.

### ➤ Select the most relevant references

For most papers, you will find many references that relate to your topic. It is important to distinguish between those related to your topic and those *relevant* to your writing. Related references may discuss your topic and may be highly interesting yet not be relevant to the arguments you are trying to make. Relevant references are those that apply to the content of your writing and to the flow of thoughts most directly and elegantly. These references are the most respected by the scientific community. Thus, rather than listing any and all papers published on the topic, select the most relevant references by citing original or review articles (i.e., articles that analyze, summarize, and critique previously published studies and findings on a common topic) and choosing the most important papers on a subject to support your document's content (this will also keep the number of your references manageable.)

One indication of the importance of a paper is the number of its citations compared to other articles on the topic. The number of citations is usually provided in the results of a literature search, and—for quick reference—can also be gleaned when doing a Google search. Another measure is the impact factor of the journal in which an article has been published. The impact factor ranks the journal based on the frequency with which average articles in it have been cited in a given year. When possible, validate specific findings. That is, use a primary source, which is the original, peer-reviewed publication of a scientist's new data, results, and theories. For a general overview of a topic, you may also use secondary sources (e.g., a review article) or certain tertiary sources (e.g., a textbook).

**➤ Verify your references against the original document**

References tend to have a surprisingly high rate of error. Therefore, when you use references found in other sources, you need to verify them against the original document. Make sure you have read all references you cite to prevent false representation of the reference or the information within. In addition, ensure that every reference in the text is included in the Reference List and that every reference in the Reference List is cited in the text. Ensure also that citations and references follow the format requested in any instructions for composing your document.

**➤ Evaluate Web sources before use**

If you are planning on using material from the Internet, evaluate the source before you use the information. If the website is that of a peer-reviewed journal, it contains primary sources, as do open access sites, that is, organizational, societal, or library databases containing full-text research articles that have been peer-reviewed, such as BioMed Central or PubMed, which provide collections of free research articles in the biological and medical sciences. For additional open access journals, see also https://doaj.org.

Many other websites may contain informative secondary sources. You should, however, verify their content and that of their citations before you use these sources. Websites that contain reliable information often have a domain extension of ".edu" (education) or ".gov" (government) or ".ac" (academic) rather than ".com" (commercial) or ".net" (Internet). Check also who created the website. Do they have expertise or credentials? Is it a reputable organization? Is their purpose clear? Moreover, find out who the intended audience is and check whether the information is current. See if the site looks professional and uses correct spelling, punctuation, and grammar. Assess if facts are represented as facts and opinions as opinions.

The following university websites have useful advice for helping students determine whether Web sources are credible:

https://usm.maine.edu/library/checklist-evaluating-web-resources
https://www.library.kent.edu/criteria-evaluating-web-resources

## 4.3 CITING REFERENCES

Whenever you use the ideas and findings of others, the source needs to be cited in the text and listed in a Reference List at the end of an article or paper. Such citations give credit to researchers for their intellectual work. They can also be used to locate specific articles, show your familiarity with the field, and help fight plagiarism.

## ➤ Know where to place references in a lab report or scientific paper

| | |
|---|---|
| **Abstract** | Do not place any citations in the Abstract. Start citing sources in your Introduction. |
| **Introduction** | Cite the most relevant references only. Although the amount of background information needed depends on the audience, do not review the literature exhaustively. |
| **Materials and Methods** | When appropriate, cite original references for methods used in your study—for example, "Growth was measured and analyzed according to Billings (1988)." |
| **Results** | Usually, statements that need to be referenced are not written in the Results section. Comparison statements are made in the Discussion section. |
| **Discussion** | Include references to compare and contrast your findings, studies that provide explanations, or those that give your findings some significance. |

## ➤ Cite references in the requested form and order

References are listed in two formats in scientific documents: as text citations and in the Reference List (or Literature Cited section). Text citations list references within the text in short version, such as by name and year. The Reference List at the end of a document displays the full citation of the reference.

In the text as well as in the Reference List, references can be cited in different ways. Some common formats for text citations are parenthetical—(author, year) and (number)—and others are bracketed or superscript or both, as shown here: "[number]." Always check the guidelines for any document you have to complete to ensure that your citations are correctly formatted. Two examples of text citations are shown in Examples 4-1a and 4-1b.

**Example 4-1** **a** Vit-E is a fat-soluble vitamin **(Hollander et al., 1976)**.
**b** Vit-E is a fat-soluble vitamin.**[8]**

If you cite multiple references for a point in your text, list the references in chronological order. If the references were published in the same year by the same author(s), add a lowercase letter after the year to distinguish the references (in alphabetical order), as in Example 4-1c.

**Example 4-1** **c** Vit-E is a fat-soluble vitamin (Traber, 1998; Brigelius-Flohé & Traber, **1999a**; Brigelius-Flohé & Traber, **1999b**).

Generally, for a publication by one author, cite that author's name.

**Example 4-2a** . . . described by Popi (18, 20).

For a paper by two authors, cite both authors' names.

**Example 4-2b** Daniles and Ebert (9) reported XYZ.

For a paper by three or more authors, cite the first author's name followed by "et al."

**Example 4-2c** . . . has previously been reported (Brown et al., 1999a; Brown et al., 1999b; Liu et al., 2003).

Note that there are a number of different style guides that specify citation format and styles for reference lists. Most instructors (and scientific journals) will indicate which style to use and provide examples of reference citations and lists for authors. If requested to use a specific style, you should follow the style's specifications. The most common reference styles and style guides in the sciences, such as that of the American Medical Association (AMA), Council of Scientific Editors (CSE; formerly CBE), or American Psychological Association (APA), are discussed in more detail in Section 4.4. AMA style is the preference for the medical sciences, whereas CSE style is used primarily in the biological and other sciences, and the APA style is used largely in psychology, social sciences, and general sciences.

### ➤ Know where to place references in a sentence

References can be incorporated into the text in two general ways. To emphasize the science, place the citation directly following a concept, idea, or finding. To emphasize the scientist, place the citation directly following the names of the author(s).

**Example 4-3**

**a** Starfish fertilization is species-specific **(17)**.

**b** Peterson **(17)** reported that starfish fertilization is species-specific.

Do not place references in the middle of an idea or after general information of a study, such as after "in a recent study" or "has been reported." Also note that references for different points in one sentence have to be cited after the appropriate point rather than grouping all the references together at the end of the sentence.

**Example 4-4** Compound A can be separated from the mixture by two methods: distillation **(Ramos et al., 2011; Smith et al., 2013)** and HPLC **(Koehler et al., 2004)**.

## 4.4 COMMON REFERENCE STYLES

### ➤ List references in the requested style in the reference list

Your Reference List at the end of your paper (commonly also referred to as "Literature Cited") should contain a list of the literature cited in the text. Many different reference styles have been developed by various scientific societies and publishing houses over the years to provide a uniform appearance of references within a field or related fields and within their respective publications. Styles differ on where to place and what to include in terms of first name initials, year of publication, journal volume, page numbers, and so on. Depending for whom your document is intended or where it is published, you will have to use the requested reference style. Often examples of the desired style are provided. Be sure to check and follow the provided reference style requirements and citation instructions carefully. Aside from determining which elements to include, check on which elements to italicize, and check on spacing and punctuation between and following elements. Follow ALL the rules to the letter.

Typically, if you have used the "(author, year)" system in the text, references are listed in alphabetical order and are not numbered in the Reference List.

**Example 4-5a**

1. Bailey, S.E., Olin, T.J., Bricka, R.M., and Adrian, D.D. (1999). A review of potentially low-cost sorbents for heavy metals. *Water Res.* **33:**2469–2479.
2. Das, N.C. and Bandyopadhyay, M. (1992). Removal of copper(II) using vermiculite. *Water Environ. Res.* **64:**852–857.
3. Hani, H. (1990). The analysis of inorganic pollutants in soil with special regard to their bioavailability. *J. Environ. Anal. Chem.* **39:**197–208.
4. Lackovic, K., Angove, M.J., Wells, J.D., and Johnson, B.B. (2004). Modelling the adsorption of Cd(II) onto goethite in the presence of citric acid. *J. Colloid Interface Sci.* **269:**37–45.

If you have used the number system in the text, references in the list are numbered in the order in which each reference is first cited in the text.

**Example 4-5b**

1. Lackovic, K., Angove, M.J., Wells, J.D., and Johnson, B.B. (2004). Modelling the adsorption of Cd(II) onto goethite in the presence of citric acid. *J. Colloid Interface Sci.* **269:**37–45.
2. Bailey, S.E., Olin, T.J., Bricka, R.M., and Adrian, D.D. (1999). A review of potentially low-cost sorbents for heavy metals. *Water Res.* **33:**2469–2479.

3. Hani, H. (1990). The analysis of inorganic pollutants in soil with special regard to their bioavailability. *J. Environ. Anal. Chem.* **39:**197–208.

4. Das, N.C., and Bandyopadhyay, M. (1992). Removal of copper(II) using vermiculite. *Water Environ. Res.* **64:**852–857.

---

References to books or book chapters also require great attention to detail. Here, too, there are various formats, and you may have to adjust yours to the one requested in class.

---

**Example 4-6** Ege, Seyhan N. (1984). *Organic Chemistry.* D.C. Heath and Company, Lexington, MA/Toronto, pp. 203–229.

---

If no citation instructions are given, consider using one of the more commonly accepted formats, such as the AMA (http://www.amamanualofstyle.com), CSE (https://writing.wisc.edu/Handbook/DocCSE.html), or APA (http://www.apastyle.org) style. Another option is to select a pertinent journal in that field and then use the format of that journal in your write-up.

---

**Example 4-7** **AMA style**

**In Text** . . . as reported previously.[13]

**Bibliography** ***Book:***

Okuda M and Okuda D. *Star Trek Chronology: The History of the Future.* New York: Pocket, 1993.

***Journal article:***

Jefferson TA, Stacey PJ, and Baird RW. A review of Killer Whale interactions with other marine mammals: predation to co-existence. *Mammal Review* 2008; 21(4):151–80.

---

**Example 4-8** **CSE style**

**In Text** . . . observed by Hinter (2008).

(McCormac and Kennedy 2004)

(Meise et al. 2003)

(Hinter 2008)

**Bibliography** ***Book:***

McCormac JS, Kennedy G. 2004. *Birds of Ohio.* Auburn (WA): Lone Pine. p. 77–78.

***Journal article:***

Meise CJ, Johnson DL, Stehlik LL, Manderson J, Shaheen P. 2003. Growth rates of juvenile Winter Flounder under varying environmental conditions. *Trans Am Fish Soc* 132(2):225–345.

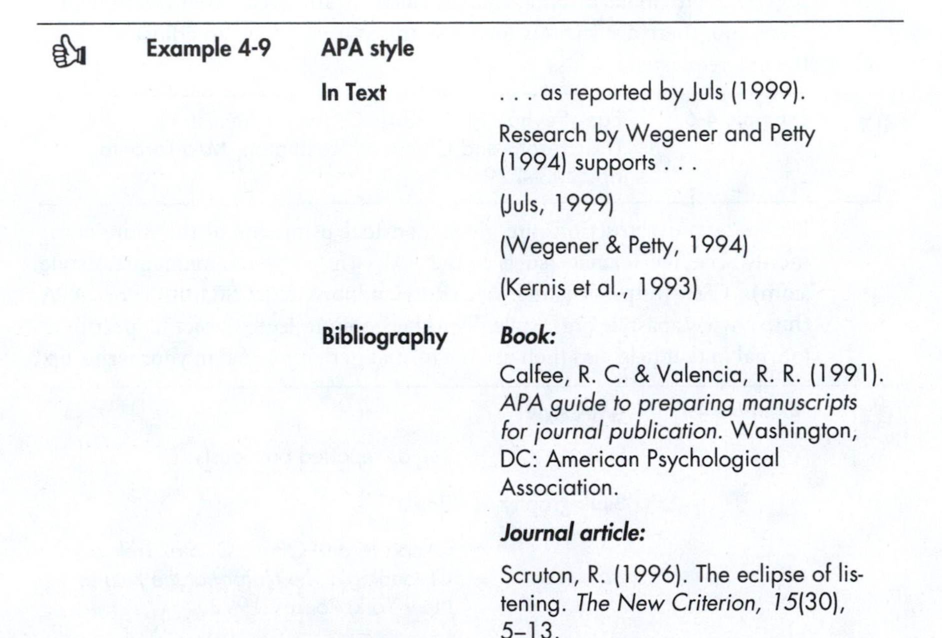

**Example 4-9** **APA style**

**In Text**

. . . as reported by Juls (1999).

Research by Wegener and Petty (1994) supports . . .

(Juls, 1999)

(Wegener & Petty, 1994)

(Kernis et al., 1993)

**Bibliography**

***Book:***

Calfee, R. C. & Valencia, R. R. (1991). *APA guide to preparing manuscripts for journal publication.* Washington, DC: American Psychological Association.

***Journal article:***

Scruton, R. (1996). The eclipse of listening. *The New Criterion, 15*(30), 5–13.

## ➤ Know how to cite and list references from the Internet

Use of Web citations is not always accepted, but this is a developing area (see also Section 4.2 and Chapter 1, Section 1.4). Generally, as the author, you can decide which style to choose. However, make sure that the style does not conflict with that asked for by your instructor.

To cite and list a reference from the Internet or the World Wide Web, use one of the following forms:

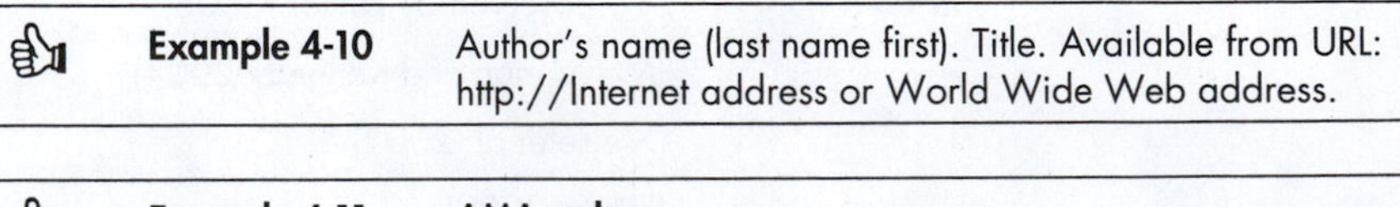

**Example 4-10** Author's name (last name first). Title. Available from URL: http://Internet address or World Wide Web address.

**Example 4-11** **AMA style**

***Online document:***

Author's name (last name first, then first and middle initial). Document title. Name of journal, if appropriate, year, volume: pages. Available at <URL>. Accessed [date]

***Book:***

Bryant PJ. The Age of Mammals. *Biodiversity and Conservation*. 28 Aug. 1999. Available at <http://darwin.bio.uci.edu/~sustain/bio65/lec02/ b65lec02.htm>. Accessed September 30, 2017.

***Article in an electronic journal (e-journal):***

Joyce M. On the Birthday of the Stranger (in Memory of John Hawkes). *Evergreen Review* 5 Mar. 1999. Available at <http://www.evergreenreview.com/102/evexcite/joyce/nojoyce.html>. Accessed November 26, 2017.

---

**Example 4-12** **CSE style**

***Online document:***

Author's or organization's name. Date of publication or last revision [year month]. Document title. Title of complete work (if relevant). <URL>. Accessed [year month].

***Book:***

Bryant P. 1999 Aug 28. Biodiversity and conservation. <http://darwin.bio.uci.edu/~sustain/bio65/index.html>. Accessed 2017 Sep 30.

***Article in an electronic journal (e-journal):***

Browning T. 1997. Embedded visuals: student design in Web spaces. *Kairos: A Journal for Teachers of Writing in Webbed Environments* 3(1). <http://english.ttu.edu/kairos/2.1/features/browning/bridge.html>. Accessed 2015 Oct 21.

---

**Example 4-13** **Chicago style**

***Online document:***

Author's name. "Title of document." Title of complete work (if relevant). Date of publication or last revision. Accessed [month year] (if required), DOI or <URL>. *(Note: A DOI, or Digital Object Identifier, is a code that identifies and organizes electronic information, including scientific literature; it can usually be found in the article information after a literature search. For more information, see http://www.biosciencewriters.com/Digital-identifiers-of-scientific-literature-PMID-PMCID-NIHMS-DOI-and-how-to-use-them.aspx)*

***Book:***

Bryant, Peter J. "The Age of Mammals," in *Biodiversity and Conservation* August 1999. <http://darwin.bio.uci.edu/~sustain/bio65/index.html> (September 30, 2017).

***Article in an electronic journal (e-journal):***

Browning, Tonya. "Embedded Visuals: Student Design in Web Spaces," *Kairos: A Journal for Teachers of Writing in Webbed Environments* 3, no. 1 (1997), <http://english.ttu.edu/kairos/2.1/features/browning /index.html> (October 21, 2015).

---

## 4.5 MANAGING SOURCES

### ➤ Manage your references well

Keep a list of references to help organize and keep track of them. There are few aspects of preparing a manuscript that are more irritating than painstakingly typing, changing, or correcting the Reference List. You can save yourself much time and much frustration if you manage your references from the start, using a reference managing computer program right when you download your references from the library or Internet. If you have many references in your list, such as when you are writing an article for publication, programs that put references in various formats (EndNote, Mendeley, or Zotero) are very useful. If you are unfamiliar with such a program, inquire at your library. Most libraries offer short classes on reference programs.

## 4.6 PLAGIARISM AND PARAPHRASING

### ➤ Ensure that you are not plagiarizing

In scientific writing, direct quotations are rarely used. Instead, information is commonly summarized and paraphrased. In all cases, the source has to be cited. Failing to indicate the source of information in scholarly scientific work is called plagiarism and is a form of academic misconduct. To give credit to the work and ideas of others, you need to acknowledge your sources, even if the writing is not absolutely identical. This rule is usually also pointed out in a school's honor code of conduct and covered during orientation.

To avoid plagiarism, you need to know what constitutes it. Plagiarism includes:

- Using material without acknowledging the source. (This is the most obvious kind of plagiarism.)
- Borrowing someone else's ideas, concepts, results, and conclusions and passing them off as your own without acknowledging them—even if these ideas have been substantially reworded.
- Summarizing and paraphrasing another's work without acknowledging the source.

The rules apply to both textual and visual information. If you are using the Internet as a source of information, you must also cite that source (see Section 4.4).

Note that you do not have to document facts that are considered common knowledge. Common knowledge is information that can be found in numerous places and is likely to be known by a lot of people, such as the information found in Example 4-14.

**Example 4-14** Many endemic species exist on the Galapagos Islands.

However, information that is not generally known (i.e., information readers outside your discipline would need to look up) and ideas that interpret facts have to be referenced as in Example 4-15 so that they are verifiable.

**Example 4-15** Based on a recent study, the blue iguanas of the Grand Cayman Islands are an endangered species (9).

The finding that "blue iguanas of the Grand Cayman Islands are an endangered species" is not a fact but an *interpretation*. Consequently, you need to cite your source to show that an actual study has been done and that this study has been accepted as fact in science. If you are uncertain whether something falls into the common knowledge category or if you have to look it up, it is best to document it.

Following are some other examples of common knowledge that do not require a citation:

**Example 4-16**

**a** As phosphorus is a key element for plant growth and is essential for many cell functions, the cycling of phosphorous in the soil has been studied widely.

**b** Volcanic eruptions are often preceded and accompanied by "volcanic unrest," providing early warning of a possible impending eruption.

Statements that contain information and interpretations that need to be cited are shown in the next examples.

**Example 4-17**

**a** Although low concentrations of phosphorus are often a limiting factor in plant growth, excess phosphorous in the soil is correlated with decreased plant health (19).

**b** The eruption of Mount Pinatubo in 1991 was preceded by a relatively short progression of precursory activity before its full-blown eruption (22).

Writing about the ideas and conclusions of others is a given in science. It is not considered plagiarism to do so as long as you acknowledge the source in your document. If you cannot verify an original source, the information should not be stated or should be clearly identified as unverified, unpublished, or an opinion (see also Chapter 1, Section 1.4).

It is easy for authors to lose track of cited and verbatim text in a larger work or document, particularly one composed over a longer period. In some cultures, the concept of plagiarism may also not exist or may be much looser than in the Western world. To verify that text is plagiarism-free, software such as Turnitin and PlagScan and apps such as Plagiarisma by Plagiarisma.net are available. These tools allow you to screen your papers for plagiarism. Be aware, though, that aside from plagiarism, other forms of ethics violations may also arise. Such ethics violations may include fabrication of data and results, fudging findings, stealing data, and being asked to include an author on a publication although the researcher did not contribute to the project. For more information on research ethics, see Chapter 1, Section 1.4.

### ➤ Know how to paraphrase

To paraphrase is to express someone else's words, thoughts, or ideas in your own words. Learning how to paraphrase is probably one of the most important skills in scientific writing. In science, you usually have to build on the work and ideas of others, but you need to paraphrase them and reference their work.

It is important that you distinguish between paraphrasing and plagiarizing. Changing a word or two in someone else's sentence or changing the sentence structure while using the original words is not paraphrasing but plagiarizing.

---

**Example 4-18 Plagiarized sentence**

*Original:*

Grizzly bears (*Ursus arctos* ssp.) encompass all living North American subspecies of the brown bear: the mainland grizzly (*Ursus arctos horribilis*), the Kodiak (*Ursus arctos middendorffi*), and the peninsular grizzly (*Ursus arctos gyas*), but none of the giant brown bear subspecies found in Russia, Northern China, and Korea.

👎 *Plagiarized sentence*:

Grizzly bears (*Ursus arctos* ssp.) consist of the North American subspecies of the brown bear, including the mainland grizzly (*Ursus arctos horribilis*), the Kodiak (*Ursus arctos middendorffi*), and the peninsular grizzly (*Ursus arctos gyas*), but not the subspecies found in Russia, Northern China, and Korea.

*Paraphrased sentence:*

Only the three North American brown bear subspecies *Ursus arctos horribilis, middendorffi*, and *gyas* are considered to belong to the grizzly bears. (Brown bears inhabiting Siberia and Northeast Asia are another subspecies.)

---

In the plagiarized sentence of Example 4-18, only a few words have been changed, omitted, or included. In the paraphrased sentence of the example, the same general idea is presented in an entirely different sentence from the original one.

Following is an example of a paragraph that instead of being paraphrased has been plagiarized:

---

**Example 4-19 Plagiarized paragraph**

*Original:*

Healthy older adults often experience mild decline in some areas of cognition. The most prominent cognitive deficits of normal aging include forgetfulness, vulnerability to distraction and other types of interference, as well as impairments in multitasking and mental flexibility (Albert, 1997; Bimonte, 2003). These cognitive functions are the domain of the prefrontal cortex, the most highly evolved part of the human brain. Prefrontal cortical cognitive abilities begin to weaken even in middle age, and are especially impaired when we are stressed. Understanding how the prefrontal cortex changes with age is a top priority for rescuing the memory and attention functions we need to survive in our fast-paced, complex culture.

*Plagiarized sample:*

In healthy older adults, often some areas of cognition decline. The most noticeable cognitive declines of normal aging include forgetfulness, vulnerability to distraction, and problems in multitasking. These cognitive tasks are localized in the prefrontal cortex, which is the most highly evolved portion of the brain. In middle age, prefrontal cortical cognitive functions already start to decrease. Such functions are also particularly affected during any type of stress. Studying memory and attention is important to understand how the prefrontal cortex changes with age. It is particularly important to understand these changes in our current fast-paced lifestyle.

---

In the plagiarized sample of Example 4-19, no sources are cited. Furthermore, only a few words have been changed in any given sentence. In addition, the sequence of sentences has been reordered, but the sentences have essentially remained the same. For each of these reasons, the derived paragraph is considered to be plagiarized.

An acceptable way of paraphrasing the preceding sample paragraph would be the following:

**Revised Example 4-19**

*Paraphrased sample:*

Studies show that the process of aging is accompanied by a decline in cognitive abilities, deficits in working memory, and compromised integrity of neural circuitry in the brain (Albert, 1997; Bimonte, 2003). If these functions of the prefrontal cortex decline, they affect our thinking and eventually our quality of life. To find ways and potential therapies to counteract this process, it is important to understand the underlying mechanisms of aging on neural circuitry.

This is acceptable paraphrasing because the writer accurately relays the information in the original using his or her own words. The writer also lets the reader know the source of the information. Following is another example of an acceptably paraphrased paragraph:

**Example 4-20**

*Original:*

Zika virus was first discovered in 1947 in the Zika Forest of Uganda.[4] It is spread largely by mosquitoes. Initially, it occurred along the equatorial belt from Africa to Asia. Starting in 2007, the virus spread to the Americas, eventually causing the 2015–16 Zika virus epidemic in South and Central America. In most adults, infection by the virus causes no or only mild symptoms. However, the virus can also be transferred from the mother to her unborn child. In these fetuses, infection by the virus can result in severe brain malformations, known as microcephaly, and other birth defects.

**Revised Example 4-20**

*Paraphrased sample:*

Zika virus is transmitted to humans mainly through mosquitoes. Its name derives from the Zika Forest of Uganda, where it was first identified in the 1940s. The virus' spread from Africa and Asia to South America led to the 2015/16 Zika virus epidemic. Although most infected adults experience comparatively mild symptoms, when the virus is transferred from mother to child *in utero*, infection can lead to microcephaly.

Unlike elsewhere in a scientific research paper, many portions of the Materials and Methods section will sound extremely similar to each other, mainly because there are only so many ways one can describe procedures whose technique and setup is essentially identical with the exception of the variables. Using very similar phrases in such passages, along with substituting your variables, would not be considered plagiarism. Therefore, do not desperately try to invent new wordings to describe the same procedure. Here are some examples of passages that would not be considered plagiarized:

**Example 4-21**

*Method description in paper A:*

Real-time fluorescence quantitative PCR was performed in an Applied Biosystems Prism 7000 instrument in the reactions containing an Applied Biosystems SYBR green master mix reagent and oligonucleotide pairs to the endogenous control gene 'A' and cDNA of 'B'. The reagents were denatured at 95°C for 10 min, followed by 40 cycles of 15 s at 95°C and 60 s at 60°C. The primer sequences (5′–3′) were 'A' forward, 5′-GACACCTATGCCGAACCGTGAA-3′; 'A' reverse, 5′-CTGAGTATCAGTCGGCCTTGAA-3′; 'B' forward, 5′-GTTCGACGACATCAACATCA-3′; 'B' reverse, 5′-TGATGACGTCCTTCTCCATG-3′.

*Method description in paper B:*

PCR amplification of 'X' sequences was done using the GC RICH PCR System (Roche, Mannheim, Germany). All non-'X' sequences were amplified using Taq DNA polymerase (Promega, Madison, WI). Primers were designed using published sequences for 'X-1' (GenBank: Xxxxxx) and 'x-13' (GenBank: Xxxxxx) (Table 1). PCR thermal cycling conditions were: 2 min at 50°C, 10 min at 95°C, followed by 40 cycles of 15 s at 95°C and 1 min at 60°C. PCR reactions were run with molecular weight standards on 0.8% agarose gels containing ethidium bromide and visualized by UV light. The primers used were: 5′-GGCTCACCAGCATCATATACG-3′ and 5′-GGCTACAATGACGACGTCA-3′.

*Method description in paper C:*

Real-time PCR was performed using the TaKaRa SYBR PCR kit and ABI Prism 7000 sequence detection system according to the manufacturer's specifications. The primers for amplification were *abc* (5′-CGCTCCTCTGCATCTAATCAG-3′ and 5′-GACACTTAGCACGCACTCA-3′) and *def* (5′-GCATCTTCAAGTAAGGACTATC-3′ and 5′-GACTTTCACAGTACCAGATT-3′). Total reaction volume was 50 μl including 25 μl SYBR Premix Ex Taq with SYBR Green I, 300 nM forward and reverse primers, and 2 μl cDNA. The thermal cycler program was 1 cycle at 95°C for 10 s, followed by 40 cycles at 95°C for 5 s and 60°C for 30 s. The PCR products were detected by electrophoresis through a 2% agarose gel stained with ethidium bromide.

For these passages and for similar ones that occur mainly in the Materials and Methods section of a research paper, it may not be a bad idea to collect sample phrases from other articles for your reference. Know, however, that I am not advising you to copy entire passages to be placed into your manuscript—only individual sample phrases and expressions that can be applied to writing your research article.

### ➤ Keep track of ideas and references

When you compose a document, you can save yourself much time and confusion if you keep track of the sources of information from the start. There is nothing more frustrating than having to identify the origin of ideas and information when you are done writing. Thus, keep a list of sources.

The best way to avoid plagiarism is to do the following when you collect and use information in scientific writing:

- Keep track of references and save the information you intend to use whenever you come across a passage that you think may be useful for your document.
- Keep a detailed list of sources.
- If you copy something word for word, put it in quotation marks, but know that writing in the sciences uses direct quotations only rarely. When you want to use details from the original but not necessarily all of them and not necessarily in the same order as the original, you need to paraphrase.
- Write the most important ideas in your own words using bullet points.
- Take notes with the book closed and without looking at the original passage on the Web. This way you are forced to put the ideas into your own words.
- Double-check that the reference and information is correct by going back to the original when you compose your document.

## SUMMARY

#### REFERENCE GUIDELINES

1. Know where to find sources.
2. Distinguish between primary, secondary, and tertiary sources.
3. Become familiar with the most important science databases.
4. Use appropriate search terms.
5. Select the most relevant references.
6. Verify your references against the original document.
7. Evaluate Web sources before use.
8. Know where to place references in a lab report or scientific paper.
9. Cite references in the requested form and order.

10. Know where to place references in a sentence.
11. List references in the requested style in the reference list.
12. Know how to cite and list references from the Internet.
13. Manage your references well.
14. Ensure that you are not plagiarizing.
15. Know how to paraphrase.
16. Keep track of ideas and references.

## PROBLEMS

### PROBLEM 4-1

**In the following paragraph, which is part of an Introduction, check that the text citations have been placed appropriately.**

Ostracodes are small bivalved Crustacea that form an important component of deep-sea meiobenthic communities along with nematodes and copepods (10). Crustaceans (10, 11) are dense and diverse in the deep sea and are one of the most representative groups of whole deep-sea benthic community. Pedersen et al. as well as Jackson et al. reported (12–14) that ostracode species have a variety of habitat and ecology preferences (e.g., infaunal, epifaunal, scavenging, and detrital feeders), representing a wide range of deep-sea soft sediment niches. Furthermore, Ostracoda is the only commonly fossilized metazoan group in deep-sea sediments. Thus, fossil ostracodes are considered to be generally representative of the broader benthic community. The distribution and abundance of deep-sea ostracode taxa in the North Atlantic Ocean are influenced by several factors (14, 15), among them, temperature, oxygen, sediment flux, and food supply. Several paleoecological studies suggest (1, 16) that these factors influence deep-sea ecosystems over orbital and millennial timescales.

*(With permission from National Academy of Sciences, U.S.A.)*

### PROBLEM 4-2

**Paraphrase the following passage.**

A drought is defined in the glossary of meteorology as "a period of abnormally dry weather sufficiently prolonged so that the lack of water causes a serious hydrologic imbalance in the affected area." Droughts can occur all over the world and at any time of the year. Although they are a frequent and often catastrophic feature in semiarid climates, droughts are less frequent and less severe in humid regions and often not even noticed in deserts. Many droughts last longer than one season and their severity and length can have dire consequences for plant, animal, and human populations. Low humidity, high temperature, and wind can aggravate drought conditions substantially, as can human water use and consumption. Consequences of a drought may include crop damage or loss, water shortage, fire risk, soil erosion, and starvation. Often these effects accrue over time and do not revert immediately after a drought ends. Drought characteristics and management are widely disputed among scientists and policy makers and therefore not much progress has been made in drought management in many parts of the world.

*(Miller J. and Carling C. 2012. Journal X, pp. 54–55.)*

**PROBLEM 4-3**

**Paraphrase the following passage.**

The highly pathogenic avian influenza A (H5N1) virus (short H5N1) was first reported in humans in Hong Kong in 1997. Humans were infected through contact with sick poultry, but outbreaks among humans have been only sporadic so far. Since 2003, 664 cases of avian influenza A (H5N1) have been reported, and 60% of the infected people have died (391 cases) (WHO, 2014). Most infections have occurred in Asia and in the Middle East. In the United States, no infections have been reported in birds or humans to date. However, scientists fear that the virus could spark a global human pandemic, which may kill millions of people. Unlike other flu viruses, H5N1 does not yet have the ability to spread between humans, but work has shown that the virus only needs five favorable mutations to become transmissible among ferrets (Herfst, 2012). Although the number of human cases has been declining since 2006, the number of outbreaks among birds and poultry remains high. In addition, the virus is constantly evolving. Therefore, WHO recommends monitoring the disease worldwide (WHO, 2014), and H5N1 vaccine has been stockpiled in the United States (CDC, 2014).

*(Smith J. (2012). Journal ABC, vol. 34.)*

**PROBLEM 4-4**

**The following passage comes from a research paper publication. Read through the original to get an understanding of its central points, and then determine whether the student revisions are plagiarized or paraphrased versions.**

**Original**

Hammerhead sharks are well known predators of fish, smaller sharks, rays, octopus, shrimp, and other crustaceans. Attacks have been observed on 32 species of cephalopods, 12 species of crustaceans, 26 species of stingrays, other hammerhead sharks, and even their own young (1). Some marine mammals, except dolphins and whales, have also been recorded as prey of hammerhead sharks (2). Hammerhead shark interactions have not been studied in detail, and research is needed to determine the importance of various prey species for hammerhead sharks. Some interactions between hammerhead sharks and other species are not predator-prey interactions, however. In some instances, hammerhead sharks have been known to harass other species and also to feed or swim alongside other marine species, ostensibly ignoring them. Some reports describe hammerheads being attacked by dolphins and killer whales. Such non-predatory interactions appear to be more common than expected. Taken together, interactions between hammerhead sharks and other species are multifaceted. More studies are needed to shed further light into their relationships and contacts.

*(Source: Haskins, T. A., Nepomucino, P., and Bert, W. (2011). ABX Review vol. 26[3].)*

**Student Version A**

Hammerhead sharks prey on other marine animals such as on squid and lobster, and on many members of the stingray family (Haskins et al., 2011). Hammerheads have also attacked members of their own family. Some interactions between hammerhead sharks and other marine species do not result in predation, for example, "harassment" by the hammerheads, feeding by different species in the same area, toleration of other species of sharks, and even attacks on hammerheads by dolphins. These nonpredatory interactions are relatively common. Thus, interactions between hammerhead sharks and marine animals are complex, and involve many different factors that we are just beginning to understand. Further work is necessary to understand the ecological interactions of hammerhead sharks.

**Student Version B**

Ecological interaction of hammerhead sharks with other marine animals can be predatory or nonpredatory. Predatory interactions involve hammerhead attacks on many marine animals. Nonpredatory interactions involve hammerheads harassing other marine animals, feeding in the same area, and even attacks on hammerheads by the other species (Haskins et al., 2011). Although these interactions have been observed, further work is needed to fully understand hammerhead shark interactions with other marine animals.

CHAPTER 5

# Basics of Statistical Analysis

THIS CHAPTER DISCUSSES:

- Nomenclature of basic statistical analysis
- How to distinguish the most common distribution curves
- How data are analyzed statistically
- How statistics are reported
- How statistics are presented graphically

A good understanding of statistics is necessary and important in the sciences. Just as students want to know the class average and high and low exam scores, scientists seek statistical data to be able to gauge the significance of findings. This chapter outlines different statistical tests and provides a general overview of basic statistics in the sciences. The chapter is not meant to serve as a detailed guide on how to apply any type of statistical analysis, however. There are many assumptions that must be understood and made for each statistical test. To gain a deeper understanding of statistics, I encourage you to take a statistics class or to read a book on statistical analysis (see Section 5.7 for suggested resources).

## 5.1 GENERAL GUIDELINES

### ➤ Use statistical analysis to determine data trends and chance occurrence

Experimental results often require statistical analysis to report overall trends and to support your interpretations and conclusions convincingly. Knowing the most important and frequently used statistical methods to analyze findings is important, particularly when you are dealing with much variation in your data. Statistical analysis enables you to judge objectively how unusual an event is. It allows you to determine the probability or likelihood of something happening randomly or because of an underlying general cause.

## 5.2 BASIC STATISTICAL TERMINOLOGY

### ➤ Know the most common statistical terms and calculations

You should be familiar with the most common terms and methods and their meanings. Note that graphing programs and spreadsheets like Microsoft Excel often have tools to perform statistical analysis (Analysis ToolPak for Excel, for example, is an add-on program that can be downloaded). These tools make it easier for students to determine statistical values.

---

***mean*** **($\bar{x}$ or $\mu$)**

The arithmetic average of a set of values, that is, the sum of the values divided by the number of the values. The mean is usually reported together with the number of measurements or sample size ($n$) and with the *standard deviation* or *range*. The mean is usually not as robust as the median, because it is influenced by extreme data points.

**Example 5-1**

Values: 2, 4, 8, 12, 16, 18

$n = 6$

$mean = (2 + 4 + 8 + 12 + 16 + 18)/6 = 10$

***median***

The middle value in a set of numbers; it separates the higher half and the lower half of the set. The median lets us understand the central tendency of a data set better than the mean, because the median is more robust.

**Example 5-2a**

Values: 2, 12, 18

$median = 12$

**Example 5-2b**

Values: 2, 12, 14, 18

There is no middle number in this set of values; therefore, the *median* = average of the two middle numbers = 13.

***mode***

The number(s) that is (are) repeated most often in a set of values.

**Example 5-3**

Values: 2, 12, 12, 18

$mode = 12$

If no numbers are repeated, there is no mode.

***range***

The difference between the maximum and minimum value.

**Example 5-4**

Values: 2, 4, 8, 12, 16, 18

range = 18 – 2 = 16

***deviation*** The difference between a data point value and the mean (also the difference between observed and expected); note: deviation ≠ *standard deviation* (see below for the definition of standard deviation).

**Example 5-5**

Values: 2, 4, 8, 12, 16, 18

$n = 6$

mean = 10

*deviation for the first data point* = 2 – 10 = –8

***variance*** **($\sigma^2$) and *standard deviation*** **($\sigma_x$)** Provides information about the variability of the data, that is, how far numbers are spread out from their average value.

The *variance* is calculated by summing up the squared deviations of each data value and dividing the sum by the number of values.

**Example 5-6a**

Values: 1, 2, 3, 4, 5

$n = 5$

mean = 3

*variance* = $((3 - 1)^2 + (3 - 2)^2 + (3 - 3)^2 + (3 - 4)^2 + (3 - 5)^2)/5 = (4 + 1 + 0 + 1 + 4)/5 = 2$

The *standard deviation* is the square root of the variance.

*standard deviation* = $\sqrt{2}$ = 1.4

***standard error*** The standard deviation of a sample divided by the square root of the sample size.

**Example 5-6b**

Values: 1, 2, 3, 4, 5

N = 5

standard deviation = 1.4

*standard error* = $1.4/\sqrt{5}$ = 0.6

***95% confidence interval*** Derived from the standard error; indicative of a 95% chance of including the true mean.

## 5.3 DISTRIBUTION CURVES

### ➤ Understand the most important distribution curves

To know what statistical tests to apply to your data, you need to have a good basic understanding of distribution curves. Only when you have determined what distribution curves your data follow can you decide which analysis to apply and what statistical values to report. The most frequently encountered distribution curves are discussed in this section.

### Binomial Distribution

The binomial distribution is the most well-known discrete distribution. The distribution is discrete because it has a limited number of potential outcomes and random variables that can be listed as whole numbers. Binomial distributions are used for samples with two possible outcomes (yes or no)—such as in a coin toss, where the outcome can be either heads or tails. Each trial, known as a Bernoulli trial, has the same *probability* ($p$), and trials are independent of each other.

The mean of a binomial distribution is $\mu = np$, and the standard deviation ($\sigma$) for binomial distributions is defined as $\sigma = \sqrt{np(1 - p)}$, where n is the number of trials. Figure 5.1 is an example of a binomial distribution.

### Poisson Distribution

The Poisson distribution is also a discrete probability distribution. It describes the probability of a given number of outcomes in a specific space, volume, or time if the average rate of occurrence is known and outcomes are independent of each other. For example, dogs have on average a litter of seven puppies. Sometimes, however, this number is smaller, and sometimes it is larger. The Poisson distribution describes the probability that the count is 1, 4, 8, or 13, or another number, for a given litter. Thus, it calculates the spread of occurrence.

When n is large and $p$ is small, the Poisson distribution approaches a binomial distribution. For large n and small $p$, the variance of a Poisson

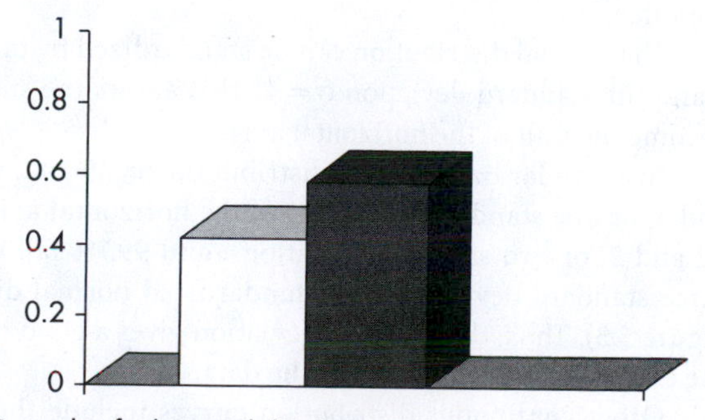


Figure 5.1 Example of a binomial distribution.

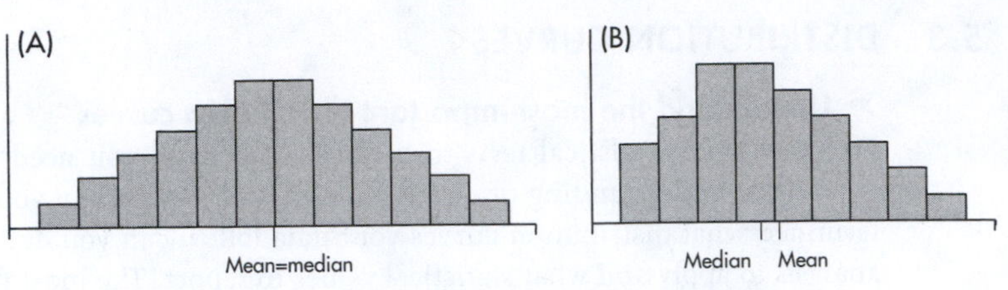


Figure 5.2 Examples of Poisson distributions: A, symmetrical; B, skewed.

random variable is the same as its mean. Figure 5.2 is an example of a Poisson distribution.

Other discrete distribution curves include the geometric distribution, the hypergeometric distribution, the negative binomial distribution, and multibinomial distributions. To learn more about these distributions, I encourage you to take a statistics class or refer to a statistics book such as *Statistics for Engineers and Scientists* by William Navidi or some of the other resources listed in Section 5.7.

### Normal Distribution and Standardized Normal Distribution

The normal distribution, also known as the Gaussian distribution, is the most important and most commonly used continuous distribution in statistics. It is continuous rather than discrete, meaning that the potential outcomes can only be described by an interval of real numbers. Continuous random variables are uncountably infinite, such as, for example, the amount of rain per year in a given location.

The normal distribution is often used in hypothesis tests to determine significance levels and confidence intervals. The distribution appears as a smooth, bell-shaped, symmetrical curve when values are graphed based on their frequency. For a normal distribution, the mean, median, and mode are identical because deviations within the data set are of equal variance. The normal distribution is thus described by only two parameters—the mean $\mu$ and the standard deviation $\sigma$. The probability for any value can be determined based on the number of standard deviations between the mean and the value.

The normal distribution can be standardized by making the mean $\mu = 0$ and the standard deviation $\sigma = 1$. This allows the standard deviation to become the unit of the horizontal axis.

In a standardized normal distribution, 68.3% of the data fall within –1 and 1, or one standard deviation, of the horizontal axis; 95.4% fall within –2 and 2, or two standard deviations; and 99.7% fall within –3 and 3, or three standard deviations (see standardized normal distribution curve in Figure 5.3). Thus, the standard deviation gives a good criterion for judging the extent of error contained in the data.

Other continuous distribution curves include the log normal distribution, the exponential distribution, the gamma distribution, and the uniform distribution. Here, too, it would be meaningful to glean details on

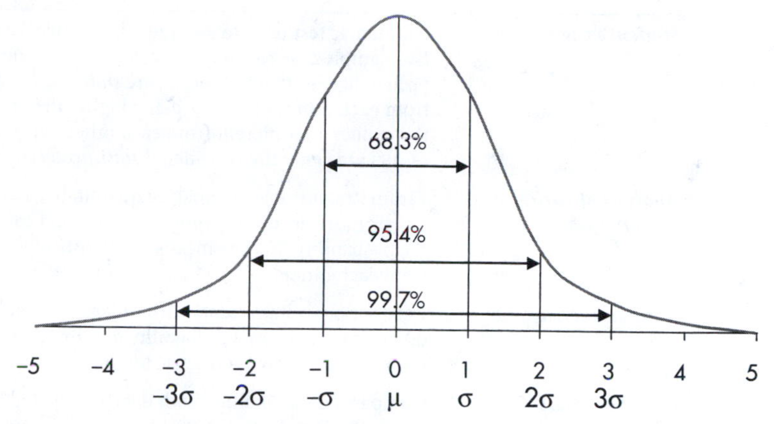


Figure 5.3 Standardized normal distribution curve.

these from a statistics book such as *Statistics for Engineers and Scientists* by William Navidi or any of the other resources listed in Section 5.7.

## 5.4 STATISTICAL ANALYSIS OF DATA

Statistics is intertwined with the scientific method. To be able to analyze your experiment statistically, you need to:

1. Design the experiment and decide on the population to be studied.
2. Collect the data.
3. Determine the correct distribution.
4. Analyze and summarize the data. Determine the mean, standard deviation, and statistical significance based on the distribution.
5. Draw conclusions to answer the research question.

### ➤ Understand the null hypothesis and statistical significance

A null hypothesis assumes that there is no difference among two or more data sets, or between results obtained and results expected. For example, in an experiment your null hypothesis might be that there is no difference between the growth rate of chives grown at 15°C and those grown at 16°C. Although you might observe a higher growth rate for 16°C than for 15°C, there is still the possibility that your results may be purely due to chance, particularly if the variability in each sample is large. To obtain a measure of likelihood of wrongly rejecting the null hypothesis, you need to conduct certain statistical tests.

### ➤ Know which statistical analysis to perform

Before conducting any parametric analysis, however, you need to determine which test is the most appropriate to use based on how well your data fit certain assumptions. An overview of the most common statistical analyses is given in the following list:

| | |
|---|---|
| ***Student's t-test*** | Parametric test used to analyze quantitative data for a two-sample case to compare means and determine the probability ($p$) that the means are *statistically* different from each other; the lower $p$, the higher the probability that values are different from each other; if $p < 0.05$, results are generally considered *statistically* significant. |
| ***Analysis of variance (ANOVA)*** | Parametric test used to analyze quantitative data for two or more groups to compare means and calculate the probability that the means are *statistically* different from each other. |
| ***Nonparametric tests*** | Used for quantitative data with different variation; data do not have a normal probability distribution; generally less powerful than parametric tests. |
| ***Chi-square test*** | Compares data obtained with data anticipated based on a hypothesis. |
| ***Regression analysis*** | Used to study the form of the relationship between two variables, such as in a dose–response relationship; in regression, one of the variables (x) is usually experimentally manipulated. |
| ***Correlation analysis*** | Determines the strength of the relationship between two variables that are measured. |

Often, scientists use a decision tree to determine which statistical test to use. The following table provides a simplified example of such a decision tree. Trees tend to be based on general guidelines that should not be construed as rules. Your data usually can be analyzed in multiple ways, and each path can

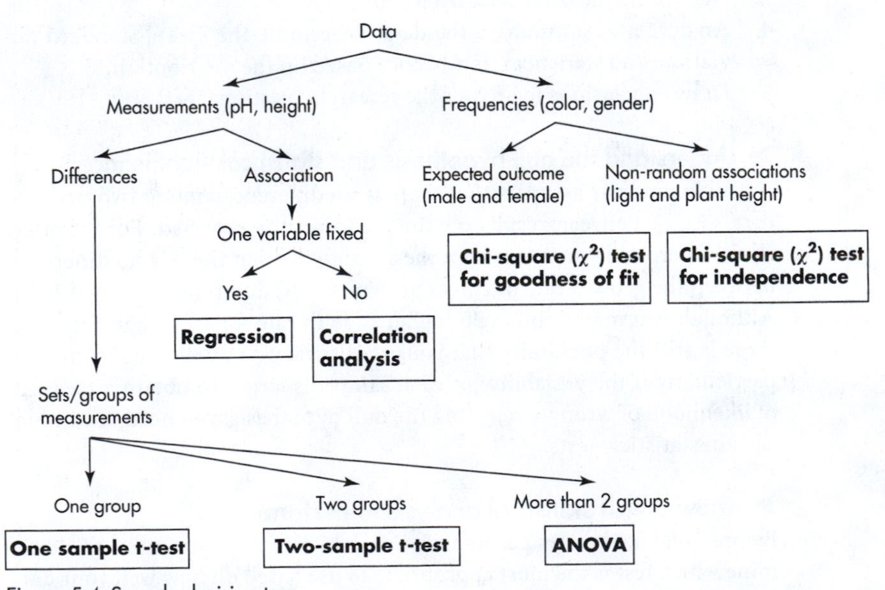


Figure 5.4 Sample decision tree.

lead to a legitimate answer. Appropriate statistical tests are selected based on the type of data, whether the data follow the normal distribution, and the overall goal of your study. For more resources, including websites, to learn about decision trees and how to analyze your data using statistical tests, see http://guides.nyu.edu/quant/choose_test_1DV and also Section 5.7.

Consider the following example, which displays the number of bacterial colonies from two different stocks growing on the same medium:

**Example 5-7**

| | **Stock 1** | **Stock 2** |
|---|---|---|
| | # colonies | # colonies |
| plate 1 | 9 | 13 |
| plate 2 | 10 | 15 |
| plate 3 | 6 | 18 |
| plate 4 | 12 | 12 |
| plate 5 | 9 | 16 |
| plate 6 | 8 | 11 |
| *mean* | *9* | *14.2* |

Although the calculated means appear quite different (9 for stock 1 and 14.2 for stock 2), suggesting that stock 2 grows more colonies on average than stock 1, it is important to confirm (or reject) this hypothesis through statistical means as the variability in each sample can be large. Confirmation can be done using a *t*-test or ANOVA. Given that you are comparing only two different groups, and no more, the simplest test to use is the *t*-test. (ANOVA would be used for more than two groups). The *t*-test gives you a statistic that you can compare in a given statistical table (along with the degrees of freedom, which can be determined based on the sample size and the number of calculations of the study) to obtain the probability or *p*-value. *P*-values below 0.05 are considered statistically significant. In our example, $p < 0.01$, and therefore the mean growth of stock 2 is significantly different from that of stock 1.

Note, though, that often *p*-values can get manipulated when data points are selected until results become significant or when various statistical analyses are performed until one is found that leads to a statistically significant result. This type of manipulation is known as *p*-hacking or "selective reporting," and essentially reports a false positive. Do not fall victim to this approach.

## ➤ Understand correlation and causation

Distinguish between correlation and causation. Many studies assess whether a correlation exists between two variables. If it does, further study may be needed to explore whether one action causes the other, that is, if there is a time order to the variables and any alternative explanation can be ruled out. For example, a scientist may find that there is a strong correlation between green geckos and temperature. This does not mean that temperature causes the geckos to be green. There may be an underlying confounder such as location. Suppose that these geckos are found primarily in the tropical rain forest where there is much green vegetation. In this case, the camouflage pigmentation of the geckos is responsible for its green color and the gecko just happens to live where it is hot. Thus, although color and temperature are correlated, temperature is not the causation for color.

### ➤ Understand how to perform statistical analysis

To conduct any statistical test properly, it is essential to take a statistics course, to read a book on statistical analysis, or to work with an actual statistician (see also Section 5.7). The tests listed above are ultimately more complicated than described here. For instance, *t*-tests can be paired (for comparing subjects or groups of data that are somehow related, such as for the same set of patients before and after treatment) or unpaired (for comparing groups of data that are not related, such as infection in group A versus infection in group B). The analysis must also consider if the data should have a Gaussian distribution or equal variance and if the *p*-value should be one-tailed (tests if the mean is significantly different from $x$ in only one direction, i.e., only one tail of the normal graph) or two-tailed (tests if the mean is significantly greater or less than $x$, i.e., it tests both tails of a normal graph). Moreover, you may have to conduct additional tests, such as post hoc tests after ANOVA (to determine relationships between population subgroups). These considerations can alter the validity of the analysis, which is often hotly contested by scientists in the literature.

## 5.5 REPORTING STATISTICS

### ➤ Know how to report statistical data

Report statistical results in the Results section of a scientific paper, in tables or their footnotes, or in figures and figure legends where data for relevant tests are presented. Do not duplicate reporting; in other words, if you report the results in a figure legend, do not repeat the information in the text, and vice versa, unless you are specifically instructed to do so.

When you report statistics, pay close attention to italics, spacing, and decimal places. Typical summary statistics that are reported include the mean and the standard deviation or the standard error of the mean. These statistical values are typically rounded to the same number of decimal places as the original data values. If original data values are integers, however, then the calculated statistical values are reported to one decimal place.

Note, though, that some *p*-values are reported to more than two decimal places. Often, the number of measurements or samples (n) or the frequency of an occurrence is also provided.

**Example 5-8a** The average height of tomato hybrids X-34 (n = 50) was 155 cm ($\sigma$ = 11.2 cm).

**Example 5-8b** The average height of tomato hybrids X-34 (n = 50) was 155 cm ± 11.2 cm.

**Example 5-8c** About 88% of tomato hybrids X-34 grew to over 150 cm, but 12% reached less than 140 cm (Table 2).

**Example 5-8d** Nearly twice as many boys (66%) than girls (34%) displayed symptoms.

**Example 5-8e** The average growth of tomato hybrids X-34 was 26.9 cm per month.

### ➤ Know how to report statistical significance

To help determine if your results are significant, support your findings with statistics (*p*-value, $\chi^2$, *t* statistic, F value, correlation coefficient, and/or regression coefficient). Understand, though, that your biological results are most important; thus, subordinate any statistical findings by placing them in parentheses.

***Significance levels/p-values*** Report *p* values either in the Results section, in a table, or in a figure legend as appropriate (e.g., "$p < 0.01$." or "$p = 0.02$"). Note that the term "significant" should be used only for statistical significance; otherwise, use "markedly" or "substantially."

**Example 5-9**

We observed a *substantial* increase in HIV infections. Disease was *significantly* delayed (10.8 years ± 2.3 years) under treatment option A ($p <$ **0.12**).

***Chi-square tests*** Report degrees of freedom and sample size in parentheses and separated by a comma following $\chi^2$; also report the *p*-value. Round numbers to two decimal places.

**Example 5-10**

There was no gender difference among infected age groups ($\chi^2(1, N = 125) = .86$, $p = 0.25$).

***Student's t-tests*** Report like chi-square values but only with the degrees of freedom in parentheses. Also report the *t*-statistic (rounded to two decimal places) and the significance level.

**Example 5-11**

Gender played a significant role in our results, $t(24) = 3.29$, $p < 0.001$, with boys displaying more symptoms than girls.

***ANOVA*** Indicate between-group and within-group degrees of freedom in parentheses and separated by comma following *F*; also report the *p*-value. Round numbers to two decimal places.

**Example 5-12**

We observed a significant effect under treatment option A, $F(1, 95) = 3.22$, $p = 0.01$, and treatment option A + C, $F(2, 98) = 5.84$, $p = 0.02$.

***Correlations*** Report the degrees of freedom (N – 2) in parentheses and the significance level. Report the correlation coefficient, r, as a number between –1 and 1 (r = 1 indicates a perfect correlation, r = –1 indicates a perfect inverse correlation, and r = 0 indicates no linear correlation between variables). Indicate also the *p*-value and possibly the sample size.

**Example 5-13**

The two variables were strongly correlated, $r(95) = 0.39$, $p < 0.05$.

***Regression*** Can be presented in the text, in a table, or in a graph. When you describe the results of your regression analysis, report the regression coefficient (β), the squared correlation coefficient ($R^2$), the F-value (F), degrees of freedom, the significance level (*p*), and the corresponding *t*-test. You may also want to report the regression equation.

**Example 5-14**

With lack of treatment, recurrence of conditions could be predicted, $\beta = -0.24$, $t(105) = 8.58$, $p < 0.01$. Lack of treatment also explained a significant proportion of variance in recurrence, $R^2 = 0.16$, $F(1, 105) = 22.44$, $p < 0.01$.

## 5.6 GRAPHICAL REPRESENTATION

### ➤ Know how to depict statistical values graphically

Statistics can also be depicted graphically, which make values visually immediately accessible for the reader. However, you need to indicate in the Materials and Methods section or figure legend what method and software package you used to obtain your statistical values. Statistical outcomes can be shown graphically in different ways.

Where appropriate, draw a vertical line to show the standard deviation ($\sigma_x$), the standard error (the standard deviation of the sample mean $\bar{x}$; standard error = standard deviation/$\sqrt{(sample\_size)}$ ), or a certain confidence interval (e.g., a 95% interval) for each data point or bar. These lines are usually drawn in pairs in line graphs, one above and one below a data point (Example 5-15). For bar graphs, it is really only necessary to draw the top line of each pair (Example 5-16). Make the lines of the error bars thinner than other lines in the main body of the data, and do not let them overlap. In the legend, tell readers what the vertical line represents, and state how many observations each mean is based on.

**Example 5-15**

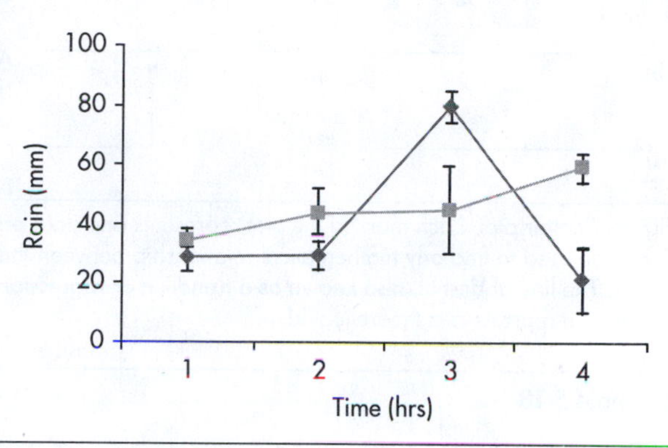


Figure. Line graph with error bars ($p < 0.5$; n = 10 for each data point).

**Example 5-16**

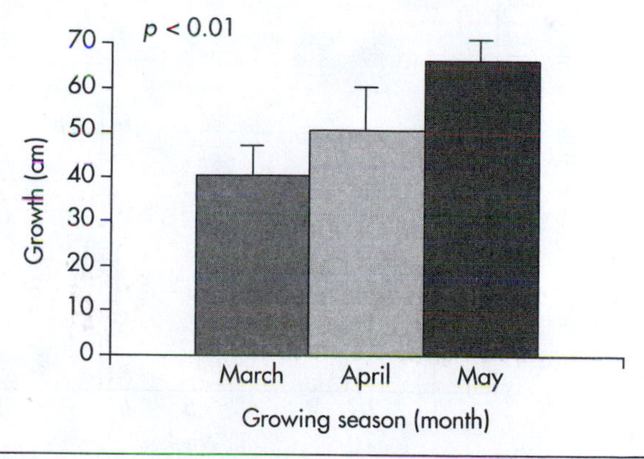


Figure. Bar graph with error bars ($p < 0.01$). Error bars typically represent one standard deviation from the mean (95% interval).

To indicate statistical values and relevance, you can also use a scatter plot (Example 5-17) or a box plot (Example 5-18).

**Example 5-17**

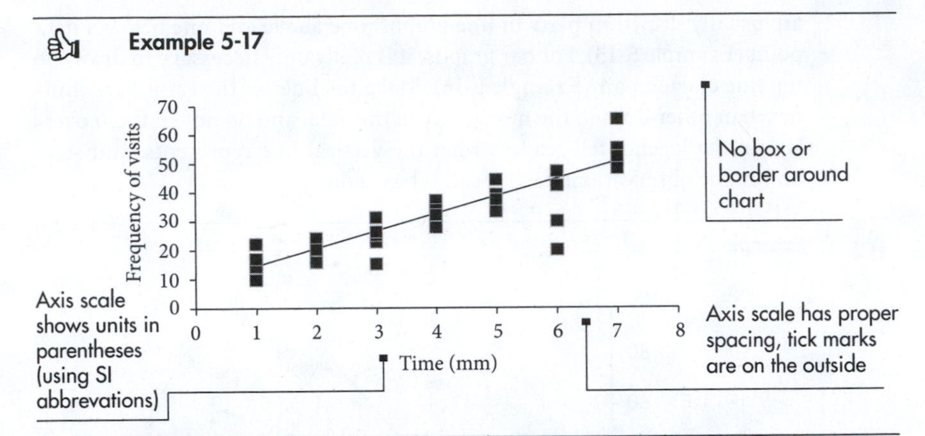


Figure. Scatter plot. Each mark in the plot represents one data point. A line of best fit can be used to find any mathematical relationship between the collection of data points. The line of best fit, also known as a trendline or regression line, can be curved or linear. It represents a theoretical ideal.

**Example 5-18**

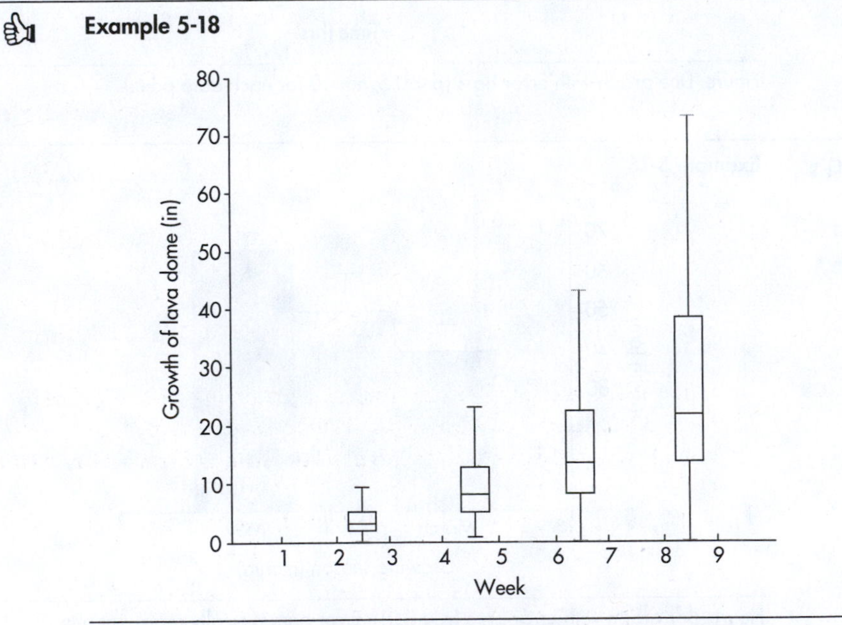


Figure. Box plot. The plot displays the median (Q2; center black line), interquartile ranges (box; Q1 and Q3), and extreme values (whiskers) from collected data.

## 5.7 USEFUL RESOURCES FOR STATISTICAL ANALYSIS

### General Statistical Analysis

- https://www.khanacademy.org/math/statistics-probability
- https://www.youtube.com/watch?v=-FtlH4svqx4

- https://www.youtube.com/watch?v=IV_m_uZOUgI
- https://www.youtube.com/watch?v=VK-rnA3-41c

## Websites for Decision Trees

- http://www.ats.ucla.edu/stat/mult_pkg/whatstat//
- http://guides.nyu.edu/quant/choose_test_1DV
- https://upload.wikimedia.org/wikipedia/commons/7/74/InferentialStatisticalDecisionMakingTrees.pdf
- http://www.csun.edu/~amarenco/Fcs%20682/When%20to%20use%20what%20test.pdf
- http://www.cbgs.k12.va.us/cbgs-document/research/Stats%20For%20Dummies.pdf
- http://www.cios.org/readbook/rmcs/ch19.pdf
- https://onlinecourses.science.psu.edu/stat500/node/68/BL

## Textbooks and Reference Books


Berry D. A., *Statistics: A Bayesian Perspective*, Wadsworth, 1996.

Bulmer M. G., *Principles of Statistics*, Dover Press, 1979.

Carlberg C., *Statistical Analysis: Microsoft Excel 2010*, Que, 2011.

Crow E. L., Davis F. A, and Maxfield M. W., *Statistics Manual*, Dover Press, 2011.

Feller W., *An Introduction to Probability Theory and Its Applications*, Wiley, 1968.

Gonick L. and Smith W., *The Cartoon Guide to Statistics*, Harper Collins, 1993.

Griffiths D. D., *Head First Statistics*, O'Reilly Media, 2008.

Huff D. and Geis I., *How to Lie with Statistics*, W. W. Norton & Company, 1993.

Milton M., *Head First Data Analysis: A Learner's Guide to Big Numbers, Statistics, and Good Decisions*, O'Reilly Media, 2009.

Navidi W., *Statistics for Engineers and Scientists*, 3rd ed., McGraw Hill, 2011.

Ott R. L. and Longnecker M. T., *An Introduction to Statistical Methods and Data Analysis*, 6th ed., Cengage Learning, 2008.

Sokal R. R. and Rohlf F. J., *Biometry*, 4th ed., Freeman, W. H. & Company, 2011.

Townend J., *Practical Statistics for Environmental and Biological Scientists*, Wiley Publishing, 2002.

Vickers A. J., *What Is a P-Value Anyway? 34 Stories to Help You Actually Understand Statistics*, Addison Wesley, 2009.

Winston W. L., *Microsoft Office Excel 2010: Data Analysis and Business Modeling*, 3rd ed., Microsoft Press, 2011.

Zar J.H., *Biostatistical Analysis*, 5th ed., Pearson, 2009.

### Computer Programs and Software

Several statistical programs exist that can be used to calculate statistical values and to visualize data graphically. Some of these are free open-source software and others are available for purchase. Add-on packages also exist for Excel. The most commonly used programs in the biological sciences include the following:

| | |
|---|---|
| **Analyse-it** | This software is largely used by scientists in the life sciences, environmental sciences, and engineering |
| **GraphPad Prism** | Originally designed for biologists in the medical sciences, GraphPad Prism is now used by throughout the biological fields as well as in the social and physical sciences. |
| **IBM SPSS Statistics** | Although originally designed for the social sciences, this package has become popular in the health sciences as well and is often used for cluster analysis. PSPP is the free open-source version of SPSS. |
| **R** | R is a free open-source statistical software language that is best used for writing custom statistical programs. It has a very steep learning curve but is highly customizable. |
| **SAS (Statistical Analysis Software)** | This software is used for statistical analysis of clinical pharmaceutical trials. The software is great for experimental design and ANOVA. The free version of this software is Dap; note that you should know basic C programming language when using it. |
| **SigmaPlot** | This scientific graphing program also incorporates data analysis. It can perform regression and correlation analysis and runs on Microsoft Windows. |
| **SimFiT** | This software is good for simulation, curve fitting, statistics, and plotting. It is very user friendly and even available in Spanish. |
| **Statistica** | This statistics and analytics package performs data mining, analysis, and visualization, and is available in several different languages. |

## 5.8 CHECKLIST

Use the following checklist to ensure that you have addressed all important elements for a figure or a table.

- ☐ When designing your experiment, did you decide on the population to be studied?
- ☐ Did you determine the correct type of distribution curve (e.g., binomial, Poisson, normal, log normal, exponential)?
- ☐ Did you determine the most appropriate statistical test to use based on your data (*t*-test, ANOVA, nonparametric tests, chi-square test, regression analysis)?
- ☐ Did you analyze and summarize the data, such as mean, standard deviation, and statistical significance based on the distribution?

- ☐ Did you subordinate statistical results in the Results section, in tables, or in figures/figure legends?
- ☐ Did you pay close attention to italics, spacing, and decimal places when reporting statistical data?
- ☐ Did you show statistical outcomes graphically?
  - ☐ Did you draw in correct error bars (lines above and below points for line graphs, lines only above points for bar graph)?
  - ☐ Did you indicate the number of observations used?

## SUMMARY

### GUIDELINES FOR STATISTICAL ANALYSIS

1. Use statistical analysis to determine data trends and chance occurrences.
2. Know the most common statistical terms and calculations.
3. Understand the most important distribution curves.
4. Understand the null hypothesis and statistical significance.
5. Know which statistical analysis to perform.
6. Understand correlation and causation.
7. Understand how to perform statistical analysis.
8. Know how to report statistical data.
9. Know how to report statistical significance.
10. Know how to depict statistical values graphically.

## PROBLEMS

### PROBLEM 5-1

**Determine the mean, variance, and standard deviation for the following data set:**

Values: 12.4, 13.5, 12.7, 14.1, 12.7, 14.2, 13.8, 20.1, 15.3, 13.6, 13.9, 14.8, 12.9, 15.0, 13.8

### PROBLEM 5-2

**For a normal distribution with a mean of 100 and standard deviation of 20, calculate:**

a) The percentage of the values between 100 and 120
b) The percentage of the values between 60 and 80
c) The percentage of the values between 80 and 140

### PROBLEM 5-3

**Ethanol is produced by yeast in media whose sugar content has an optimal concentration of 5 mg/ml. At concentrations higher than 8 mg/ml, the yeast dies, and the brewing process has to be suspended. If the sugar**

**concentration in the medium is normally distributed with a standard deviation of 1 mg/ml, what percent of the time will the brewing process need to be suspended?**

**PROBLEM 5-4**

**During an experiment you determine that bacteria X generation times have a normal distribution with a mean of 20 min and a standard deviation of 5 min. What percent of bacteria have a generation time of more than 30 min? Less than 15 min?**

**PROBLEM 5-5**

**Are the following statistics reported accurately in a biological lab report or research article? Correct the reporting as needed.**

a) We determined that the average sea surface temperature was 24°C ± 2°C in July.
b) For $n = 12$ and $\sigma = 2$, we determined a mean run rate of 11.5 cm/hr on the thin layer chromatography for compound X.

**PROBLEM 5-6**

**You have repeatedly measured the rate of oxygen production for leaf segments from the plant *Arabidopsis thaliana*, and you have found that oxygen was produced at a rate of 12 bubbles/min when the leaf segment was 5 cm from the light source and at a rate of 10 bubbles/min when the leaf segment was 20 cm from the light source. How would you describe these findings in a sentence in the following cases?**

a) Your calculations indicate that $p = 0.05$.
b) Your calculations indicate that $p = 0.75$.

CHAPTER 6

# Notebooks, Data, Figures, and Tables

THIS CHAPTER DISCUSSES:

- What information is in a lab notebook
- When to use figures, tables, or text to present data
- How to design figures and tables
- Where to place information in figures and tables
- How to highlight data
- Different types of figures
- The content of figure legends

## 6.1 KEEPING A LABORATORY NOTEBOOK

### ➤ Treat notes as records

An accurate record of your procedures and data in form of a laboratory notebook is a key component in science. Your lab notebook should be a complete record of what you did, observed, and concluded for your experiments. It should be detailed enough to allow another scientist to reproduce your work.

In the professional world, the notebook is also a legal document, which can be scrutinized in the case of a patent, for example, or in claims of fraud. It should not contain theoretical musings or notes on communications with others. The notebook, which belongs to the institution, serves as a reference for members of the lab long after you have moved on. You should feel free, however, to copy or duplicate your notes and take those copies with you.

When you are required to keep a notebook as part of a course, you may be allowed to record lecture notes, pre-lab notes, questions, and the like in your notebook. You may also be required to note the title of the lab study, and to enter an introduction, purpose/goal/hypothesis/objective, calculations, list of reagents, and a summary in addition to the detailed procedures you perform and the data you obtain.

## ➤ Organize your notebook

### Number, Date, and Sign Each Page

Entries to your lab notebook should be clear and legible. Unless pages are numbered already, enter the page number and date on each page you recorded information for an experiment. Do not write on the back of pages in notebooks that include duplicate copies. Use titles and informative headings as needed to help locate specific information.

If you need to correct a piece of information in your notebook, strike a single line through the mistake and write the correct information next to it. Do not erase notes or remove any original pages. If you left any blank spaces, put a diagonal line through these to void them to ensure no information is added later on. Sign your pages at the end of an experiment and have your instructor countersign them as well. Make the appropriate entry for the experiment with date and page numbers in the table of contents section at the front of your notebook. This way, you and others can find the needed information in your notebook quickly.

### Know What to Include in a Laboratory Notebook

After the title of the experiment, explain the goal or objective and include necessary background and references as appropriate. Then, record all procedures and observations. When you describe procedures, ensure that you also record all relevant information including temperature, amount of time and reagents, source of reagents (manufacturer and lot number), dilutions, G-force, instrument type, and so on for all procedures. If you are following a published protocol, note any changes or alterations you make to the protocol—planned or not. If you are repeating a procedure previously described in your notebook, refer back to the appropriate procedure and page. You do not need to record the exact same protocol again.

The laboratory notebook usually contains much more information than what will be provided in the corresponding research paper. For example, you may record which type of cuvette or spectrometer you used in your notebook, but would not specify these in the actual research paper unless they are essential for the success of the experiment.

Record all observations carefully, including control results and failed results. Include data records or indicate where these are stored and how they were analyzed. Include calculations and mention specific software if needed

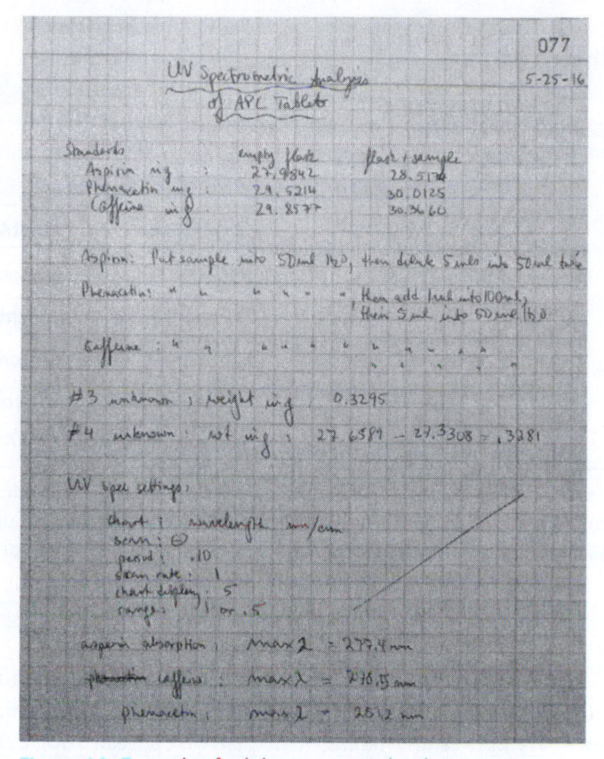


Figure 6.1 Example of a laboratory notebook page.

(see Figure 6.1 for an example). When you finish your experiment, write a brief conclusion as well as a recommendation for next steps if appropriate (e.g., what changes to make if the experiment is repeated) at the end.

## 6.2 FROM NOTEBOOK ENTRIES TO PRESENTING DATA

### ➤ Decide whether to present data in an illustration or in the text

Illustrations, such as figures, tables, photographs, and diagrams, are important components of most scientific documents and presentations. They are meant to present data visually. However, designing the illustrations is usually more time-consuming than describing results in the text, and may not be the best way to present the data to the reader. Therefore, decide first on the best way to show your data—that is, in an illustration or in the text.

For example, if your proposed table has only one or two rows of data, consider presenting your findings in one or two sentences in the text instead of constructing a table. Similarly, if your table lists descriptions in words rather than numbers, consider whether you really need a table—a few sentences in the text may be better. The table in the next example could easily be converted to text.

**Example 6-1**

**Antibiotic Targeting of Various Organisms**

| ANTIBIOTICS | ORGANISMS | CELLULAR TARGET |
|---|---|---|
| I | *Bacillus* | **ribosomes** |
| I | *Saccharomyces* | **mitochondria** |
| II | *Bacillus* | **ribosomes** |
| III | *Streptococcus* | **cell walls** |
| III | *Saccharomyces* | **mitochondria** |

**Revised Example 6-1**

*Bacillus* ribosomes were targeted by antibiotics I and II, *Streptococcus* cell walls were targeted by antibiotic III, and *Saccharomyces* mitochondria were targeted by antibiotics I and III.

When you have chosen to use an illustration rather than text, you will need to decide what format to use. The two most common formats in the sciences are figures and tables. It helps if you spread your data out on the table and arrange them in all possible combinations. Look for patterns to help you decide on what type of format to present your data.

### ➤ Present data in graphs when trends or relationships need to be revealed

Generally, choose figures when trends or relationships are more important than exact values or when hidden relationships or trends need to be revealed. Graphs are often needed to analyze data using statistical programs. Furthermore, when your graph includes the standard deviation (SD or $\sigma_x$) or the standard error (SE) (see Chapter 5), it allows readers to comprehend the magnitude of statistical differences at a glance.

### ➤ Prepare tables rather than graphs when it is important to give precise numbers

Choose tables to report precise numerical information or to compare component groups, or when data are not enough to produce a satisfactory graph. Tables present data more precisely than a graph but usually do not clearly show trends within your data. Tables also present facts more concisely than text does while allowing a side-by-side comparison of the content.

Consider the next two illustrations. Which of the following presentations would you prefer given the same data?

**Example 6-2a**

**Sample Data Presentation A**

| TIME (DAYS) | HORMONE A | HORMONE D54 |
|---|---|---|
| 0 | 200.5 | 455.8 |
| 5 | 187.1 | 356.7 |
| 10 | 166.5 | 321.9 |
| 15 | 201.1 | 400.6 |
| 20 | 289.8 | 500.7 |
| 25 | 204.1 | 489.9 |
| 30 | 189.9 | 389.4 |
| 35 | 288.9 | 513.4 |
| 40 | 205.1 | 499.3 |
| 45 | 182.9 | 298.5 |
| 50 | 278.8 | 533.2 |
| 55 | 223.4 | 498.5 |
| 60 | 199.6 | 250.6 |

**Example 6-2b**

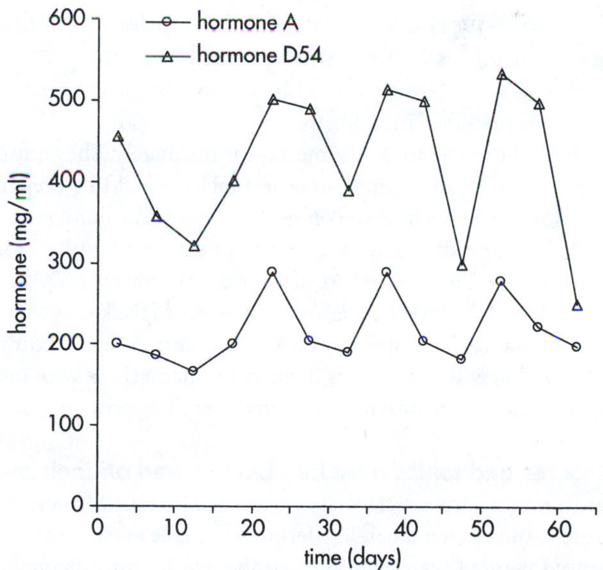


Figure. Sample data presentation B. The same data as in Example 6-2a are presented as a line graph instead of a table.

Most readers would prefer Example 6-2b because the trend and relationship of the data are more obvious, and exact numbers seem not as important.

Consider another example: The data shown in the bar graph in Example 6-3 would best be described in the text because there are only two data points, and exact values seem important.

**Example 6-3**

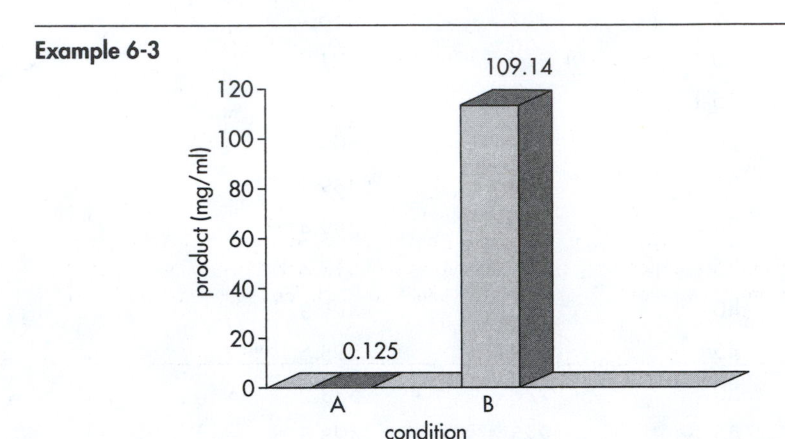


Figure. Bar graph for which data could be presented in the text instead.

## 6.3 FROM DATA TO RESULTS

### ➤ Prepare figures and tables with the reader in mind; design figures and tables to have strong visual impact

In the same way that text should be organized for the reader, figures and tables should be formatted for the reader. Figures and tables must be clear enough for the reader to get the message immediately. They must also be visually appealing. In addition, figures and tables should be prepared following instructions that may be given on specific styles. The main purpose of figures and tables is to visualize data and to support your results. Not all the data need to be described in the text of your paper, however. Describe only your key findings in the text. Use figures and tables to back up your conclusions.

Note that readers typically only pay attention to a maximum of four to five items. You will lose readers if you overwhelm them with too many items in figures and tables, or with too many overall figures and tables.

### ➤ Figures and tables must be able to stand on their own

Figures and their legends, as well as tables and their titles, must be independent of the text and of each other. Readers must be able to understand what is being portrayed without searching through the text for an explanation. The figure in Example 6-4 and its legend, for example, stand on their own. Together they provide sufficient information for the reader to understand what is being shown. If the legend or parts of it were missing, this would not be the case.

**Example 6-4**

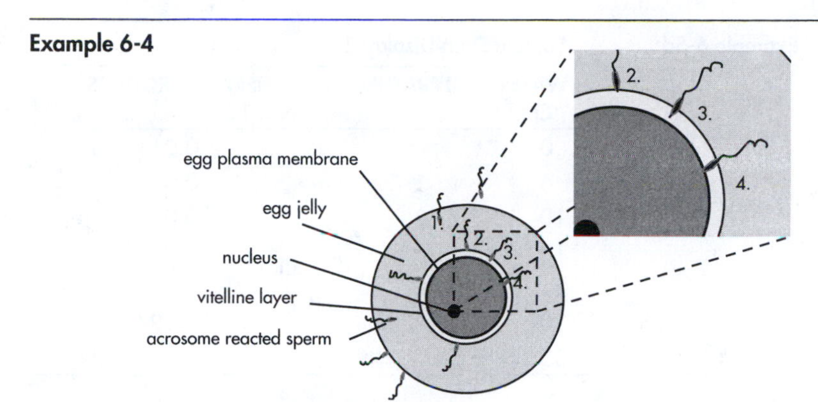


Figure X. Diagram of S. *purpuratus* egg and sperm fertilization. 1. The egg triggers the sperm acrosome reaction. 2. The sperm attaches to the egg vitellin layer at the acrosomal process. 3. The acrosomal process extends as it penetrates the vitellin layer. 4. The plasma membranes of the egg and sperm fuse.

## ➤ Place information where the reader expects to find it

Most readers can understand the intended meaning of what is presented only if the illustration has been formatted for this interpretation. To this end, scientists can present data in different listings and table formats. See Example 6-5, including the tables under Example 6-5b, c, and d, for various possibilities for a set of data in a table.

**Example 6-5a**

0°C, 0.011% hermaph-
rodites 25°C, 51/1,000
water *T* = 30°C, 0.124%
35°C, 0.152%

6°C, 0.011% herm.
18°C, 0.028%
*T* = 10°C , 2/100

**Example 6-5b**

**Sample Data Display 1**

| % HERMAPHRODITES, N = 120 | WATER TEMPERATURE (°C) |
|---|---|
| 0.011 | 0 |
| 0.011 | 6 |
| 0.020 | 10 |
| 0.028 | 18 |
| 0.051 | 25 |
| 0.124 | 30 |
| 0.152 | 35 |

**Example 6-5c**

**Sample Data Display 2**

| Water temperature (°C): | 0 | 6 | 10 | 18 | 25 | 30 | 35 |
|---|---|---|---|---|---|---|---|
| % Hermaphrodites: | 0.011 | 0.011 | 0.020 | 0.028 | .051 | .124 | .152 |

**Example 6-5d** **Sample Data Display 3**

| WATER TEMPERATURE (°C) | % HERMAPHRODITES, N = 120 |
|---|---|
| 0 | 0.011 |
| 6 | 0.011 |
| 10 | 0.020 |
| 18 | 0.028 |
| 25 | 0.051 |
| 30 | 0.124 |
| 35 | 0.152 |

Although the exact same information appears in all formats, most readers prefer Example 6-5d because it is the easiest to interpret. The reason for the easier interpretation is twofold: (a) The data are written as a table, and, even more important, (b) this table is structured such that the familiar context (temperature) appears on the left, whereas the interesting results appear on the right in a less obvious pattern. Familiar context is usually denoted as the independent variable (i.e., the variable that you control/change in order to observe or measure changes in the dependent variable).

Usually, information in a table is more readily available for readers than information in the text or information presented as raw data as in Example 6-5a. However, some tables are easier to interpret than others. Generally, readers find tables much harder to follow if they are presented as shown in Example 6-5b, or if the table is horizontally arranged as in Example 6-5c. That is, we interpret information more easily if it is placed where readers expect to find it. (Note that an alternative presentation of this data could also be in a bar or line graph, which readers would prefer even more than a well-designed table.)

### ➤ Know where to place figures and tables in professional manuscripts

With very few exceptions, professional scientific journals require that figures, figure legends, and tables are placed at the end of the manuscript when it is submitted. Figure legends can be combined in one separate section, but figures and tables themselves are typically each placed on its own page before you submit your manuscript. Check also *Instructions for Authors* to ensure that illustrations have been placed correctly and in the format requested.

## 6.4 FIGURES, THEIR LEGENDS, AND THEIR TITLES

When you prepare a figure, consider what kind of figure you need. You can choose to present your data in a **photograph**, a **drawing**, a **diagram (i.e., an explanatory image or sketch)**, or a **graph**. Most common in the sciences are graphs, which can be presented as line graphs, bar graphs, or scatter plots.

### ➤ Use line graphs for dynamic comparisons

Line graphs are the graphs most commonly encountered in science. They use points connected by lines to show how something changes in value. Line graphs display continuous data as points joined by a line, such as data that change over time. Continuous data are usually associated with some sort of physical measurement—typically, data that fall within a range and are fractions or decimals. In a line graph, multiple data sets can be graphed together, but a key must be used.

Do not place too much information into one figure—it will appear too packed. Do not leave much white space either—the graph will appear not well-constructed, and editors are not fond of too much open space that could be used otherwise. Three or four curves should be the maximum in a line graph, especially if the lines cross each other two or more times. When curves must cross, use lines of different thickness or patterns. Draw straight lines between data points or use best-fit, smoothed curves. For an example of a line graph, see Example 6–6.

**Example 6-6 Sample line graph**

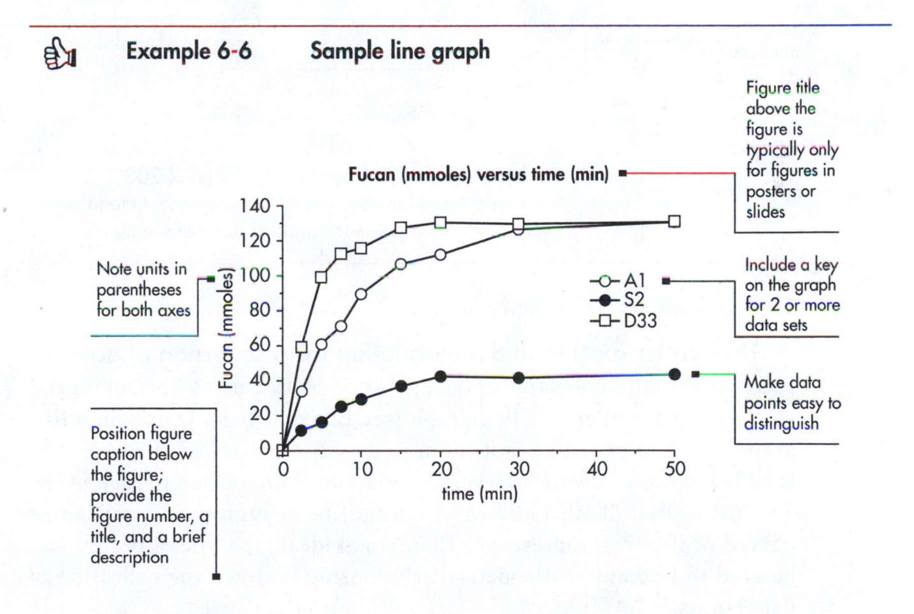


Figure. Well-organized line graph. Curves and data points are easily distinguishable. Data stand out well, and axes are designed and labeled clearly.

### ➤ Use bar graphs to compare or display findings that do not change continuously

Another common graph in science is the bar graph, which uses parallel bars of varying lengths to display comparative values (see Example 6-7). Use bar graphs instead of line graphs for discrete data (data that can be counted in whole numbers), to compare or display data that do not change continuously, and for findings that can be subdivided and compared in different ways. Bar graphs tend to be more effective than line graphs for general audiences. If you use a bar graph, use a vertical rather than a horizontal bar graph because most readers are accustomed to the former. Keep the widths of the bars, and the space between bars, consistent.

**Example 6-7 Sample bar graph**

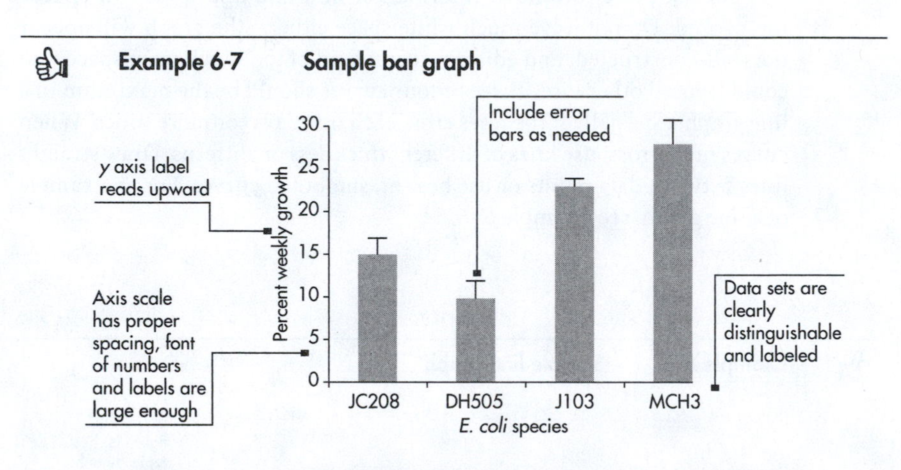


Figure Y. Percent weekly growth for *E. coli* species JC208, DH505, J103, and MCH3 in medium A. Bars represent mean (n = 10), and error bars represent standard error of the mean.

### ➤ Use scatter plots to find a correlation for a collection of data

In a scatter plot, a collection of data points is plotted using horizontal and vertical axes, similar to a line graph (see Example 6-8). Unlike in a line graph, the data points are not connected by a line. Instead, a best-fit line is added to show how the two variables ($x$ and $y$) are related to each other. The line of best fit, also known as a trendline or regression line, can be curved or linear. It represents a theoretical ideal. The line of best fit can be used to find any mathematical relationship between the collection of data points.

Data points in scatter plots often overlap. That is fine. A scatter plot allows you to find a clear correlation for the points, such as a linear relationship as shown in this example:

**Example 6-8 Sample scatter plot**

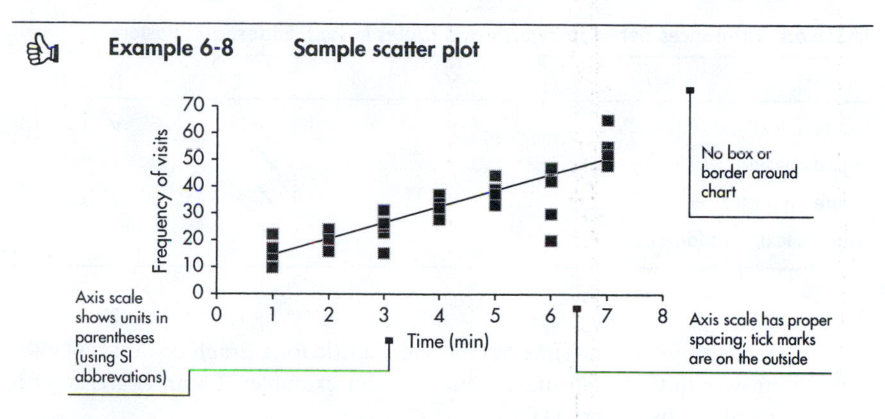


Figure. Well-presented scatter plot. Data are easy to distinguish, and axes are designed and labeled well.

## ➤ Plot the independent variable on the *x* axis and the dependent variable on the *y* axis

Readers expect to see information in figures at certain places. For line graphs and bar graphs, readers expect to find the independent variable on the *x* (or horizontal) axis and the dependent variable on the *y* (vertical) axis. If information is plotted as expected, readers can follow it more easily and do not have to spend extra time trying to understand illustrations. An example of a well-constructed line graph is shown in Example 6-6.

## ➤ Distinguish between figures and tables in text, slides, and posters

Figures and tables in text are not the same as those on slides or on posters. Distinguish between them. They differ mainly in the use of an overall title and figure legend (see Table 6.1.) Although some lab reports may require you to add a title to your figure, this is usually not the case in professional scientific articles. In professional writing, titles for figures do not appear as a header of the figure itself. Figures and tables used in oral presentations or on posters, in contrast, usually have a title/header (see the following table and also Chapters 13 and 14). These figure titles should be brief and written as an incomplete sentence without abbreviations. Often, figure titles also state important results depicted by the figure.

If you are required to add a title to your figure, the easiest way to do so is by describing what is graphed in terms of *y* axis versus *x* axis as in Example 6-6; for more advanced writing, a title for the figure is usually moved into the figure legend. Often this title is somewhat more descriptive (for example: "Aggregation of fucan for different species of sea urchins" instead of "graph of fucan aggregation versus time"). Because graphs usually

**TABLE 6.1 Differences Between Figures and Tables in Text, Slides, and Posters**

| | TITLE | LEGEND |
|---|---|---|
| Figure in text | – | ✓ |
| Figure on slide | ✓ | – |
| Figure on poster | ✓ | ✓ |
| Table in text, on slide or poster | ✓ | – |

result from an experimental question, a title for a graph could also be the answer to the experimental question (for example: "Fucan aggregation is regulated by A1 and D3").

### ➤ The figure legend should be a description of the figure content

A figure legend, or caption, is typically provided close to each figure in a scientific document. The figure legend should objectively describe the contents of the figure. Include enough information to allow the reader to understand the figure easily. A figure and its legend should also be able to stand alone. The legend should contain:

- A title
- Description of figure contents
- Explanations of symbols and abbreviations

Like the titles appearing on figure of slides and posters, those found in figure legends should be brief. They are written as incomplete sentences without abbreviations. In addition to the title, description of content, and explanation of symbols, you should provide statistical and experimental details as appropriate and required. Make sure you follow instructions given to you for your figures and their legends. Some instructors request only a title, whereas others require experimental details in the legend.

The descriptive material in a legend may include letters or symbols to explain special abbreviations and symbols shown in the figure. Abbreviations and symbols should be consistent between the figure and its legend, consistent between legends, and consistent with the text as well. Do not refer to the text for explanation. Note the concise listing of symbols in Example 6-9.

**Example 6-9**

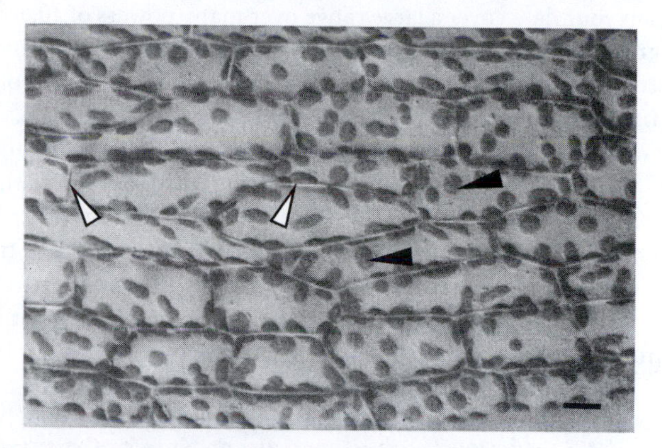

Figure. Light micrograph of *Physcomitrella patens* leaf section. White arrows, cell wall; black arrows, chloroplasts. Bar = 10 μm.

### ➤ Make each figure easy to read; differentiate points, lines, and curves well; label axes and scales well

For good first impressions, make each figure easy to read (see Example 6-6). Above all, make the key information immediately obvious. Highlight the most important information through different colors, line weights, arrows, call-outs, font sizes, or labels. At the same time, deemphasize less important information such as axes, axis labels, keys, titles, and figure labels. In addition, label axis and scales well. In particular:

- Make the lettering and symbols big enough to be legible and discernible *after* the graph is reduced.
- For lettering, use the font type Arial or Arial Narrow no smaller than size 8 to allow readability.
- Use short, informative descriptions to label the axes.
- Use the same coordinates for different figures if values in them are to be compared.
- For symbols, do not use X, +, 0, or *—these symbols are not distinctive enough. Instead, go with the most common symbols, such as circles, triangles, and squares.
- Use the same symbols when the same entities occur in several figures.
- If possible, keep similarly shaped and colored symbols separate.
- Do not clutter your graphs. Avoid the use of grid lines within the graphs. Grid lines are useful when plotting points but only rarely afterward.
- Do not extend axis lines beyond the last marked scale point.
- Do not end axes with an arrow pointing away from zero.

Where appropriate, draw a vertical line to show the standard deviation, the standard error, or a certain confidence interval (e.g., a 95% interval) for each data point (see also Chapter 5, Section 5.6).

### ➤ Know how to graph on a log scale

When your data cover a large range over many powers of 10, it makes sense to graph it using logarithmic scale. This approach creates a more uniform distribution of data points. In log base 10 each successive whole number is 10 times larger than the preceding whole number.

Semilog graphs use a logarithmic scale on one axis (usually the *y* axis) and a linear scale on the other axis (usually the *x* axis). Use these types of graphs if the range of one variable is huge and that of the other is not. If you were to convert the values of the variable with large range to their log, you could also plot the graph on two linear scales. The next figure lets you see how an exponential curve in linear space becomes a straight line in logarithmic space:

On the logarithmic scale, each cycle division differs by a factor of 10 (or a multiple thereof). So, for data ranging from 1 to 1,000, label the first tick mark as 1 in the first cycle, the one after that as 2, then 3 and so forth until you reach 10, which is the start of the next cycle. The tick mark after

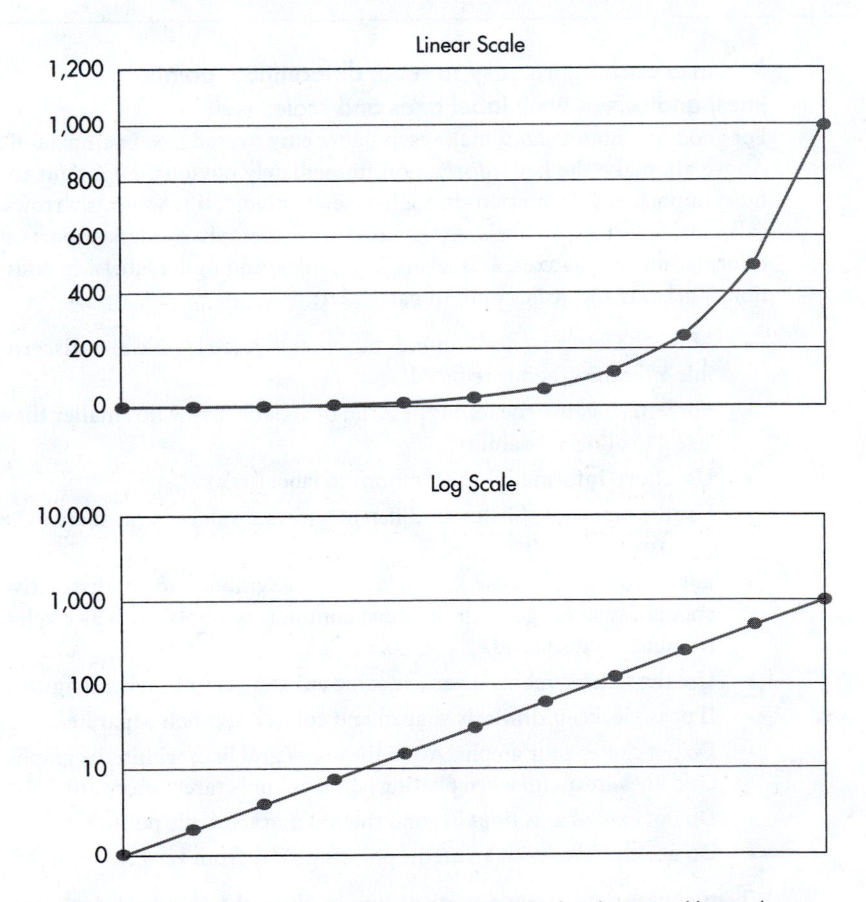


Figure 6.2 Linear versus log scale. Identical data graphed on linear and log scale.

10 would then be labeled as 20, followed by 30, and so forth until you reach 100 in the third cycle. The third cycle would go to 1,000 in multiples of 100 (see Example 6-11).

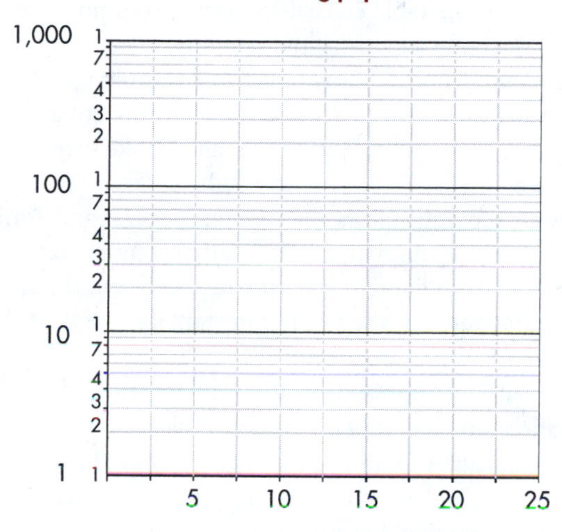


For data with much lower values, you could also start with 0.001, then the next cycle would be 0.01, then 0.1 and 1.0; for larger data values, you could use $10^3$, $10^4$, $10^5$, and $10^6$. Note that the horizontal distances remain the same for each of the logarithmic sections.

## ➤ Do not mislead readers

Constructing a graph is more than just plotting points. It requires that you understand the rationale behind the quantitative methods. It also requires that you are as objective and honest as possible. Remember that the data you present may be interpreted in more than one way. You can highlight, exaggerate, or even misrepresent a given set of data, depending on the way you choose to display it.

You should make readers aware of certain trends without misleading them. It may be very tempting to distort information and trends by, for example, deleting data points or by massaging line fits. Resist these temptations.

Statistical information will lend validity to your data. Use error bars (see Example 6-7), or show variability of data points through scatter plots (Example 6-8). Indicate the statistical test(s) used, the number of times an experiment was repeated, the sample size, *p*-values, and other statistical information.

Often, readers are misled when different scales are used to compare data in different graphs. For such comparisons, ensure identical scales so readers can compare data directly, and arrange graphs next to each other to help assessment. In addition, do not extrapolate a line graph past the last data point, as such curves can be misleading.

## 6.5 TABLES AND THEIR TITLES

### ➤ Keep the structure of your table as simple as possible

A typical scientific table consists of a title, column headings, row (or side) headings, the body (the rows and columns containing the data), and, usually, explanatory notes. Many tables in scientific papers follow the pattern shown in Examples 6-12a and 6-12b, with row entries, column headings, and the body of the table. Within the table, make sure that your listings and numbers are aligned.

Design separate tables for separate topics, and limit the number of tables. Do not use tables just to show off how much data you have collected. Also, do not repeat data in tables if you plan to put the same data in the text or in a figure, but state the most important data in the text as well.

### ➤ Place familiar context on the left and new, important information on the right

Like figures, tables exhibit data visually, and the information presented should be independent of the text. In your tables, arrange the data to allow for easy interpretation. Because readers read from left to right, they interpret tables more easily if you place familiar information (independent variable) on the left and new, important information (dependent variable) on the right. In addition, organize your tables logically to make mental comparisons easy for readers (e.g., pretreatment or control measurements should precede posttreatment or experimental measurements). It is also easier to follow similar components when they are arranged vertically, not horizontally.

### ➤ Design table titles to identify the specific topic

The titles of your tables precede the tables and identify what is being displayed. The titles have to be specific and informative without being wordy (see Examples 6-12a and 6-12b). Avoid nonstandard abbreviations.

### ➤ Label dependent variables in column headings and independent variables in row headings

Columns and rows in a table typically have a heading. Column headings should label dependent variables, and row headings should label independent variables. Columns and rows may also have subheadings. Headings and subheadings both should include units in parentheses if appropriate.

Set the column and row headings apart from the rest of the table by using a different font, bold face, or capitalization. To keep column headings brief, and thus save space in the table, use short terms or abbreviations in the column headings and subheadings, and explain the abbreviations in footnotes if necessary.

Capitalize column headings the same as the table title, but use a smaller font. Center one-line column headings, and write all headings horizontally. If headings are too long, turn them 45 degrees counterclockwise.

Keep row headings short as well, and left justify or right justify them. Capitalize only the first letter of the first word in these headings.

**Example 6-11a General format of a table**

Position table caption above the table and provide a title

**Table 1 Initial Stocking of *R. sylvatica, A. maculatum, and A. opacum* in Pond Enclosures**

Headings include units in parenthesis where needed

| SPECIES | POND 1 ENCLOSURE[a] (NUMBER OF INDIVIDUALS) | POND 2 ENCLOSURE (NUMBER OF INDIVIDUALS) |
|---|---|---|
| *Rana sylvatica* | 100 | 50 |
| *Ambystoma maculatum* | 50 | 100 |
| *Ambystoma opacum* | 10 | 10 |

[a]Footnote *a*.

**Example 6-11b Another general format of a table**

**Table 1 Colony Diameter of *E. coli, S. aureus, and S. pneumonia* Under Different Antibiotic Media Additions**

| | AMPICILLIN (5 mg/l) | RIFAMPIN (10 mg/l) | STREPTOMYCIN (10 mg/l) |
|---|---|---|---|
| Diameter of *E. coli* (mm) | 0.21 | 0.12 | 0.43 |
| Diameter of *S. aureus* (mm) | 0.45 | 0.16 | 0.56 |
| Diameter of *S. pneumonia* (mm) | 0.25 | 0.96 | 0 |

## 6.6 SOFTWARE RESOURCES FOR FIGURES AND TABLES

Constructing good tables and figures requires the skills of specific software programs such as Microsoft Excel. Tables and figures can easily be prepared electronically. For example, you can use spreadsheets such as Excel to convert your data into a table. Alternatively, you can use Microsoft Word or other word processing software to create a table directly in a Word file (see also Appendix B). Note, however, that Excel is usually not sufficient to create figures and tables for professional journals and needs to be used in combination with other programs.

For the top 20 tips for using Excel, see Appendix C. Excellent tutorials for Excel are readily available online through various university sites and YouTube. Many also allow you to practice your skills at the same time. Useful Web addresses for tutorials include (last accessed February 2018):

**http://excelcentral.com/**—Contains online courses and video tutorials for beginning, intermediate and advanced learners.

**http://www.improveyourexcel.com/category/excel-features/**—Has free videos that are geared mostly toward the intermediate user.

**https://www.udemy.com/courses/search/?q=excel&siteID**—On Udemy, Excel courses are offered for free for all levels.

**https://www.youtube.com/user/learnexcelfunctions**—Great YouTube channels for beginning users of Excel.

**http://office.microsoft.com/en-us/training-FX101782702.aspx**—Contains links to free training courses and tutorials for diverse Microsoft products, including Excel.

**http://www.lynda.com/Excel-tutorials/Excel-2013-Essential-Training/116478-2.html**—These Excel video tutorials teach core and advanced features and tools of Excel.

**http://www.gcflearnfree.org/excel2013**—Contains lessons, interactives, videos, and apps for Excel.

**http://www.ncsu.edu/labwrite/res/gt/gt-menu.html**—Walks you through graphing with Excel.

## 6.7 CHECKLIST

Use the following checklist to ensure that you have addressed all important elements for a figure or a table.

### Lab notes

- ☐ Did you number each page consecutively?
- ☐ Did you date and sign the pages?
- ☐ Did you void all blank portions?
- ☐ Did you include a goal/objective of the experiment?
- ☐ Did you record all procedure steps and data?
- ☐ Did you include all relevant details for the procedures?
- ☐ Did you include a conclusion/next step section for the experiment?
- ☐ Did you include calculations?
- ☐ Did you make an entry into the table of contents?

### Illustrations

- ☐ 1. Did you choose the best illustration format for your data?
  - ☐ a. Did you choose text if data can easily be explained there?
  - ☐ b. Did you choose a line graph for dynamic comparisons?
  - ☐ c. Did you choose a bar graph for findings that can be subdivided and compared?

☐ 2. Can your illustration and its legend stand on their own?
☐ 3. Did you place information where the reader expects to find it (independent variable on the *x* axis, dependent variable on the *y* axis)?
☐ 4. Did you place information at the right place within the manuscript?
☐ 5. Are your figures easy to read?
  ☐ a. Are points differentiated well?
  ☐ b. Is there a limited number of curves on your graphs?
  ☐ c. Is the lettering large enough?
  ☐ d. Has the key been added to the figure?
  ☐ e. Has a size marker been added for micrographs or other pictures?
  ☐ f. Did you label axes well and include units in parentheses?
  ☐ g. Are tick marks facing outward and are there not too many?
  ☐ h. Did you use the same scales for data that is meant to be compared?
☐ 6. Did you include statistical data if needed?
☐ 7. Are all abbreviations explained in the legend?

### Tables

☐ 1. Is the structure of your table as simple as possible?
☐ 2. Did you place familiar context/independent variable on the left and new, important information/dependent variable on the right?
☐ 3. Does your table title identify the specific topic?
☐ 4. Did you label dependent variables in column headings and independent variables in row headings?
☐ 5. Did you include footnotes where necessary?
☐ 6. Are all abbreviations explained?

## SUMMARY

### GUIDELINES FOR THE LAB NOTEBOOK

1. Treat notes as records
2. Organize your notebook
3. Number, date, and sign each page
4. Know what to include in a laboratory notebook

### GENERAL ILLUSTRATION GUIDELINES

1. Decide whether to present data in an illustration or in the text.
2. Present data in graphs when trends or relationships need to be revealed.

3. Prepare tables rather than graphs when it is important to give precise numbers.
4. Prepare figures and tables with the reader in mind.
5. Design figures and tables to have strong visual impact.
6. Figures and tables should be able to stand on their own.
7. Place information where the reader expects to find it.
8. Know where to place figures and tables in professional manuscripts.
9. Use line graphs for dynamic comparisons.
10. Use bar graphs to compare or display findings that do not change continuously, and when the findings can be subdivided and compared.
11. Use scatter plots to find a correlation for a collection of data.
12. Place the independent variable on the *x* axis and the dependent variable on the *y* axis.
13. Distinguish between figures and tables in text, slides, and posters
14. The figure legend should be a description of the figure content.
15. Make each figure easy to read.
16. Differentiate points, lines, and curves well.
17. Label axes and scales well.
18. Do not mislead readers.

### GUIDELINES FOR TABLES

1. Keep the structure of your table as simple as possible.
2. Place familiar context/independent variable on the left and new, important information/dependent variable on the right.
3. Design table titles to identify the specific topic.
4. Label dependent variables in column headings and independent variables in row headings.

## PROBLEMS

### PROBLEM 6-1 Figure or Table?

**What format (table, graph, and/or text) would you use to present the following examples in a report or research article? Justify your choices.**

1. You have studied ice location, thickness, and consistency in Greenland, and have collected a series of data on these. What is the best way to present this data?
2. Your lab has assessed the rate of mortality for different mammals exposed to the West Nile virus in all the counties of Connecticut. What type of table or figure should you use to present your findings?
3. You and your colleagues have assessed the effectiveness of a new Lyme disease vaccine after inoculating mice. You have evaluated changes in their leucocyte and platelet counts with various doses of the vaccine over six months. During the course of this study, you have also taken photographs of the animals and the vivarium. What information should you present in a paper?

4. You plan to publish a paper on a new species of fungus you discovered during a trip to the rainforest in Costa Rica. You have several nice photographs of yourself in the rainforest as well as photographs of four known species of fungus and one of the new species. Which photograph should you include in your publication?
5. You want to explain a new type of apparatus, a rotating cylindrical annulus apparatus used in your experiments. Should you use a schematic or describe it in the text or both?
6. You have constructed a new drug delivery system involving the use of nanoparticles to deliver anticancer agents to specific brain tumors. Should you explain the system using a schematic or describe it in the text?

**PROBLEM 6-2 Graph or Table**

**Given the raw data provided (in triplicate readings), create a table or figure, whichever represents the data best. Time is in hours; unit of measurement is in milliliters.**

$$\text{Time 0, KA} = \begin{cases} 6.7 \text{ (1 reading)} \\ 8.1 \text{ (2 reading)} \\ 7.0 \text{ (3 reading)} \end{cases} \quad \text{KB} = \begin{cases} 3.9 \\ 3.9 \\ 3.8 \end{cases} \quad \text{TFR} = \begin{cases} 0 \\ 0 \\ 0.1 \end{cases}$$

$$\text{Time 2, KA} = \begin{cases} 7.2 \\ 7.4 \\ 6.9 \end{cases} \quad \text{KB} = \begin{cases} 5.0 \\ 5.1 \\ 4.7 \end{cases} \quad \text{TFR} = \begin{cases} 1.2 \\ 2.1 \\ 1.4 \end{cases}$$

$$\text{Time 3, KA} = \begin{cases} 5.8 \\ 6.7 \\ 6.1 \end{cases} \quad \text{KB} = \begin{cases} 5.4 \\ 5.5 \\ 5.3 \end{cases} \quad \text{TFR} = \begin{cases} 1.4 \\ 1.0 \\ 1.0 \end{cases}$$

$$\text{Time 4, KA} = \begin{cases} 5.8 \\ 6.0 \\ 5.7 \end{cases} \quad \text{KB} = \begin{cases} 5.9 \\ 5.9 \\ 5.8 \end{cases} \quad \text{TFR} = \begin{cases} 2.0 \\ 1.0 \\ 1.1 \end{cases}$$

**PROBLEM 6-3 Graph or Table**

**Given the raw data provided, create a table or figure, whichever represents the data best.**

Time 0:

Population A: 123 members
Population B: 54 members
Population C: 99 members

Time 1 month later:
Population A: 133 members
Population B: 28 members
Population C: 98 members

Time 2 months later:
Population A: 142 members
Population B: 6 members
Population C: 98 members

**PROBLEM 6-4**

**Evaluate the graph in the following figure. How could this graph be improved? Make a list.**

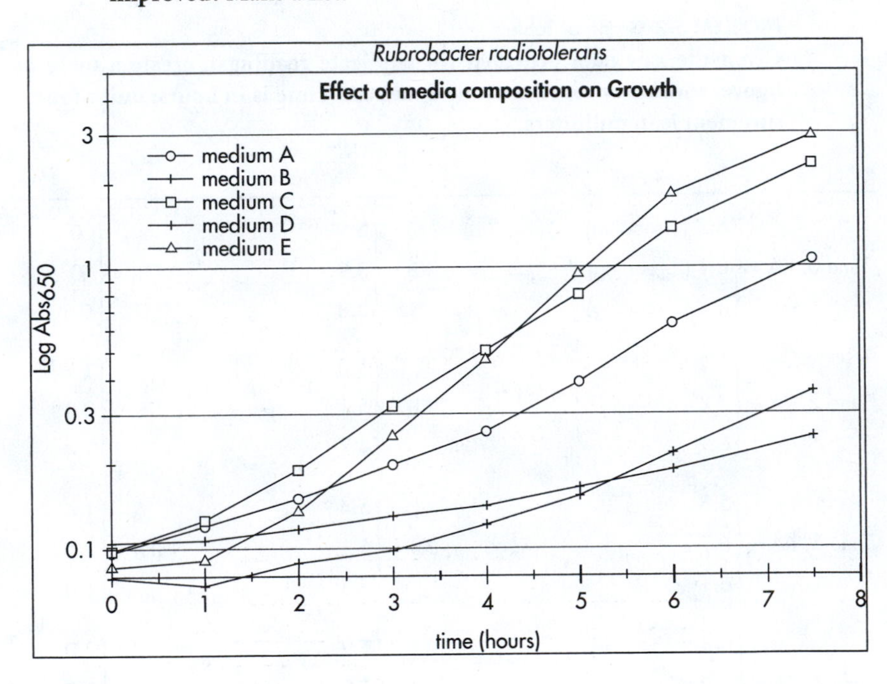

**PROBLEM 6-5**

**Evaluate the following figure. How could this figure be improved? Make a list.**

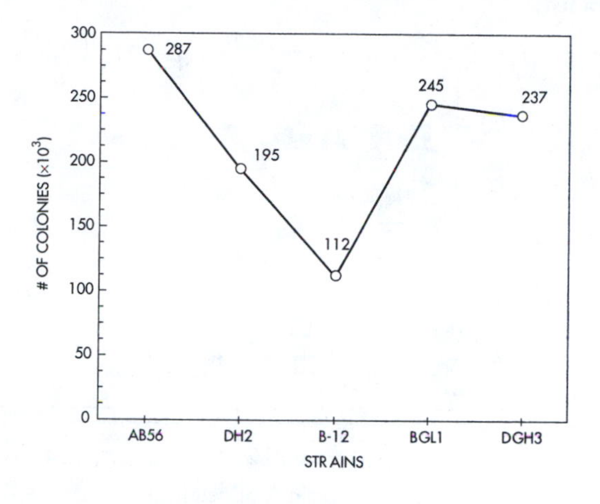


**PROBLEM 6-6**

**Evaluate the following figure. How could this graph be improved? Make a list.**

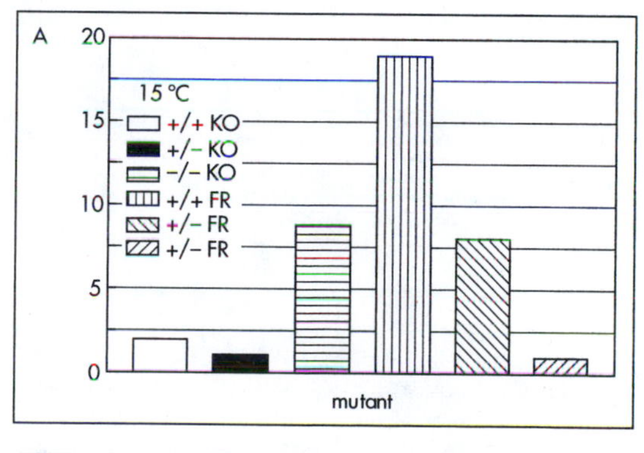


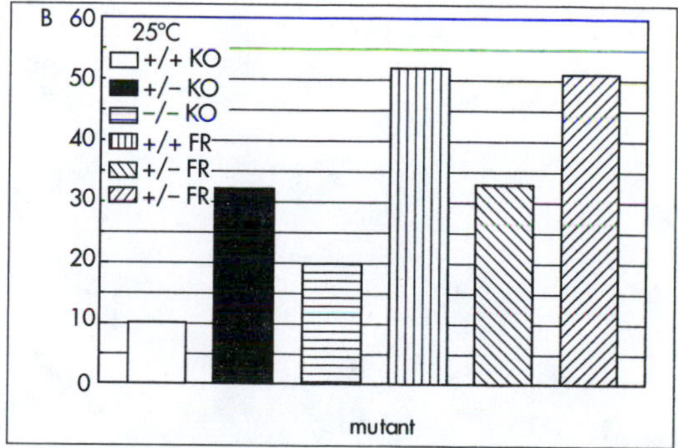

**PROBLEM 6-7**

**Evaluate the following figure. Why would readers object to this graph? Make a list.**

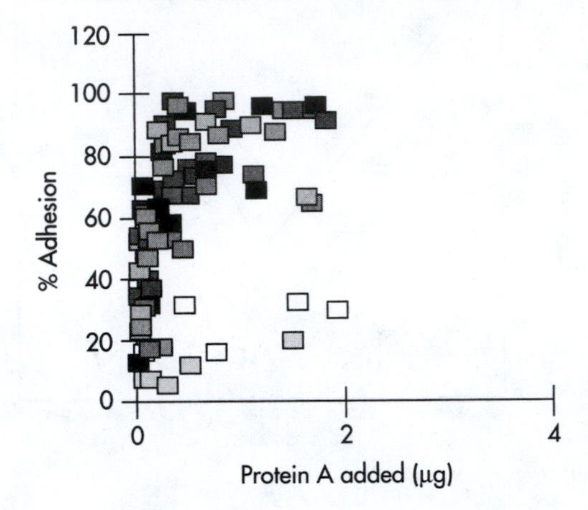

PART III

# Introductory Writing

CHAPTER 7

# Laboratory Reports and Research Papers

THIS CHAPTER INCLUDES:

- The components of a laboratory report and those of a research paper
- The IMRAD format
- Format, elements, and signals for the different lab report and research paper sections
- Annotated examples for all sections and components

## 7.1 GENERAL PURPOSE

Communicating your findings makes them available to others, and serves as an indication of your expertise and productivity. One way to learn how to communicate your scientific findings is through lab reports, the most commonly encountered document in undergraduate science courses. Although lab reports can contribute significantly to your overall grade in a lab course, the purpose of learning to write a good lab report is not only to let your professor evaluate you as a student. The exercise also serves as practice for writing a full scientific research article as a future professional. It is with this foresight that this chapter describes how to write both lab reports and scientific research articles, also known as "research papers."

## 7.2 THE BIG PICTURE

### ➤ Understand the "big picture"

In contrast to lab reports, scientific research papers present original research findings in academic journals. Whereas lab reports are typically short reports on a single experiment, small set of experiments, or study performed as part of a course, research papers tell a longer story on a whole family of related, original experiments or observations performed over months or years by professional scientists to prove or disprove a hypothesis or report details on a new finding. Research papers should not be confused with term papers you may be asked to compose when writing a literature review in your class. (Note that scientists also publish papers that review literature; these are known as "review articles" or "review papers" and are described in more detail in Chapter 11.)

This chapter describes the basic elements of a lab report, and thus lays the foundation for composing future scientific research papers. The chapter also discusses additional details needed for writing a full-fledged research paper under each section. Even if you do not become a scientist, these skills are useful as you gain a better understanding of how to read and write academic publications and acquire the scientific reasoning process.

## 7.3 BEFORE YOU GET STARTED

Before writing your first draft, you need to understand the study to be undertaken. This includes taking good notes and being aware of the protocol to follow. It also includes writing down your observations. In addition, you have to know what format to follow to write up your report, and you may need to look at diverse sources to find the references to cite in your report.

### ➤ Understand what is involved

To write the best report, you need to:

- Understand the procedure, why you are doing it, what you hope to learn, and why we benefit from this knowledge.
- Take good notes during any lectures on the material.
- Record and analyze the data meticulously.
- Think about the best way to organize the data.
- Target your write-up to a peer. Do not assume that the instructor knows everything already.

### ➤ Obtain instructions, and follow them

Ensure that you have instructions on how to write your lab report and that you understand them. Read these instructions. Note important details such as length of abstract, sections/headings of an article, writing style, format of references, and electronic format. Follow the instructions!

### ➤ Collect, organize, and study your references

To write your report, you may need references. Ask your instructor for references and sources if needed. For more information on references, see Chapter 4.

## 7.4 COMPONENTS AND FORMAT

### ➤ Follow the IMRAD format

Usually, your instructor will provide you with an outline of what is expected as a lab report for the class you are taking. The components and format of such a report may vary depending on your instructor and the class. Regardless of variations, however, your lab report should contain Introduction, Methods, Results, and Discussion (IMRAD) sections; this reflects the order of the core sections in academic journal articles and research papers:

IMRAD

**I**ntroduction
**M**aterials and Methods
**R**esults
**D**iscussion

Aside from the core IMRAD sections, usually an abstract is included. Some lab reports may not requests this, but professional research papers always do. Also typically included is a title page, references, figures, and tables. Sometimes, the Discussion section is referred to as the "Conclusions" section, and occasionally the order of sections is changed (for example, the Methods section may appear last). Overall, however, the IMRAD format represents a textual version of the scientific method: developing a hypothesis, testing it, and deciding whether your findings support the hypothesis. Reports and papers that do not follow this format, or differ significantly from it, disorient scientists. Understanding the overall format of scientific research papers and the underlying purpose of each section within them will let you adapt this overall format to the needs of a lab report for a particular course or professor.

Each section within the IMRAD format contains certain elements. To help you understand what information will go into these sections, Table 7.1 provides an overview of them. Lab instructions or journal guidelines should also be consulted as a source of information about what to include.

### ➤ Focus on the overall question or hypothesis of the study

Both lab reports and research papers should be centered on the overall question or hypothesis of the study. A good lab report or research paper also makes apparent the writer's understanding of the concepts behind the data. Do not merely present data of expected and observed results in your report or paper. You need to show your understanding and knowledge of the principles behind the study and discuss how and why differences were observed. Thus, you need to organize your ideas carefully and express them coherently. A full-length, annotated lab report is presented in Chapter 9.

**TABLE 7.1 Overview of Purpose and Content for Main IMRAD Sections**

| SECTION | MAIN PURPOSE | DETAILS |
|---|---|---|
| Introduction | States your overall question or hypothesis | Provides context by explaining how the research question was derived and how it connects to previous research; gives the purpose of the experiment/study; does not include figures or tables |
| Methods | Describes your experimental approach | Describes how you tested the hypothesis and why you performed your study in that particular way; rarely includes figures or tables, and if so, only those pertaining to study set-up |
| Results | Presents the data collected | Explains how the data were evaluated; shows calculations; contains most tables and figures to support the results |
| Discussion | Considers whether your data support the hypothesis | Interprets your results and discusses the implications of your findings; states potential limitations of your study design; sometimes includes additional figures and tables such as a new model |

## 7.5 TITLE AND TITLE PAGE

### ➤ Make the title succinct, clear, and complete

For research papers, the title is the single most important phrase of an article because most readers will judge the paper's relevance on the title alone. You may or may not be asked to include a title for your lab report. If you need to include a title, state the main topic in twelve to fifteen words and make the title interesting to attract readers.

A strong title should fulfill three criteria: It needs to be succinct, clear, and complete. A title is typically stated as a phrase, but it may also be a complete sentence. However, a full sentence with an active verb is usually not a good title—neither is an overly long or catchy phrase.

### ➤ Avoid nonstandard abbreviations

Avoid abbreviations in titles unless they are well-known abbreviations, acronyms, or contractions, or are accepted conventions, such as chemical formulas. Examples of such abbreviations include the following: $H_2O$, $CO_2$, ATP, DNA, RNA, PCR, spp., and so forth. Also avoid phrases such as "A study of . . . "—these only add bulk.

Unclear titles confuse or mislead readers and can give an annoying first impression. Some titles are not clear because word choice is too general, as shown in Example 7-1.

---

👎 **Example 7-1** Effect of light on cacti

---

This title is not clear because words in it are unspecific, such as the category terms “light” and “cacti.” What light and which cacti? The specific type of light and cacti should be listed. Otherwise, the title is essentially meaningless. At the same time, this title is unclear because it does not state what specific effect has been observed in the study. The revised example is a much stronger title:

| **Revised Example 7-1** | Effect of ultraviolet light on the growth and drought resistance of saguaro cacti |
|---|---|

Announcing the main variables of the paper is stronger than trying to fit all the variables into the title. Concentrate on the most distinctive aspect of your work. Keep in mind that in the title, readers can only absorb three or four details and that the title cannot replace the Abstract. However, be sure that your title is complete.

To check that your title is complete, compare your title with the question or hypothesis of the paper and ensure that you use the same main key terms in the title as in the question or hypothesis of the paper. Consider the following title:

| **Example 7-2** | Isolation of the fungus *Aureobasidium pullulans* |
|---|---|

Although this title orients the reader to the area of research, it does not give any specifics about where the organism was isolated. Adding a few more specific words completes the title and sets it apart from others in the field.

| **Revised Example 7-2** | Isolation of the fungus *Aureobasidium pullulans* from the vascular plant *Psilotum nudum* |
|---|---|

### ➤ Format the title page correctly

Not all lab reports have title pages, but if your instructor wants one, a typical title page may contain the following:

- The title of the study
- Your name and the names of any lab partner(s)
- Your instructor’s name
- The date the lab was performed or the date the report was submitted

**Example 7-3 Sample Title Page**

**Analysis of restriction endonuclease digests of plasmid pET3 by agarose gel electrophoresis**

Natasha Richards
Lab partner: Amy Fernández

Biology 106
Instructor: John Soders
January 25, 2018

## 7.6 ABSTRACT

### ➤ Include the following elements in the Abstract:

- Question/purpose
- Experimental/study approach
- Results
- Conclusion (answer)/implication
- Optional: short background and significance

The title page is usually followed by the Abstract. Knowing how to write an Abstract is one of the most important skills in science because virtually all of a scientist's work will be judged first (and often last) based on an abstract. The ability to write a competitive abstract applies not only to lab reports but also to other critical scientific documents such as research papers, grant proposals, progress reports, project summaries, and conference submissions. Therefore, learning this critical skill cannot be underestimated.

The Abstract should fully summarize the contents of the report or paper in one paragraph. It must also be written such that it can stand on its own without the text. Include only the most important details of the paper, and use as few words as possible. Abstracts for lab reports are typically 50 to 150 words long. Those of academic research papers are usually longer and range from 100 to 250 words.

### ➤ Do not include general overview sentences, nonstandard abbreviations, references, figures, or tables

Although the Abstract is a mini version of the report, it does not give equal weight to all the parts of a report. The Abstract may include a sentence or two of background information. It needs to include the overall question or purpose of the work. It typically describes the experimental/study approach only generally and includes only the main results from the Results section. In the Discussion, it also contains the answer to the research question/purpose. In addition, the Abstract may end with a sentence stating an implication, a speculation, or a recommendation based on the answer. It should not end, however, with a general descriptive statement that merely hints at your results. Therefore, do NOT include vague generalities such as "Results are discussed" or "A new model is presented."

In addition, in your Abstract, do not include any information or conclusion not covered in the paper. Avoid abbreviations, unfamiliar terms, and citations. Do not include or refer to tables or figures, nor any references, but be sure to include all the important key terms found in the title because the Abstract and the title have to correspond to each other. You may want to wait to write your Abstract until you have completed all other parts of your report because the Abstract will contain the main elements of all the other sections.

### ➤ Signal the elements of the Abstract

Because Abstracts are usually written as one paragraph, it helps the reader if you signal the different parts of the Abstract. Examples of signals for the abstract are shown in Table 7.2.

**TABLE 7.2 Signals for the Abstract**

| QUESTION + EXPERIMENT/ OBSERVATION | RESULTS | ANSWER/ CONCLUSION | IMPLICATION |
|---|---|---|---|
| To determine whether . . . , we . . .<br>We asked whether . . .<br>To answer this question, we . . .<br>X was studied by . . . | We found . . .<br>Our results show . . .<br>Here we report . . . | We conclude that . . .<br>Thus, . . .<br>These results indicate that . . . | These results suggest that . . .<br>These results may play a role in . . .<br>Y can be used to . . . |

Example 7-4 is an example of a well-written abstract that contains all the necessary elements.

**Example 7-4** **Sample Abstract**

| | |
|---|---|
| Hypothesis | We tested the hypothesis that an increase in temperature results in faster development of tadpoles into frogs. |
| Experimental Approach | In our experiment, bullfrog tadpoles (*Rana catesbeiana*) were kept in either 10°C, 15°C, 20°C, or 25°C warm water, and their rate of development was evaluated by scoring the appearance and growth rate of limbs over a period of four weeks. |
| Results | We observed that tadpoles living in warmer water developed limbs faster than those living in cooler water. |
| Conclusion | These results suggest that tadpoles living in warmer climates or warmer ponds might turn into frogs sooner due to more favorable living conditions. |

## 7.7 INTRODUCTION

**➤ Follow a "funnel" structure for the Introduction and include:**

- Background
- Unknown/problem
- Question/hypothesis
- Experimental/study approach
- Optional: results/conclusion significance

The purpose of the Introduction is twofold: to interest your audience to read the paper and to provide sufficient context or background information for readers to understand why your study was performed and what specific research question or hypothesis you addressed.

Introductions for lab reports are usually short—some are as short as a single paragraph—stating the question/purpose or hypothesis of the study and the approach. In addition, they may contain some background information.

**Example 7-5** **Sample Introduction for a lab report**

| | |
|---|---|
| Purpose | In this experiment we tested the effect of different water sources on tobacco pollen germination. |
| Experimental Approach | *Nicotiana tabacum* L. var. Petit Havana (tobacco) was suspended in germination medium prepared with either tap water as delivered to the laboratory, pond water collected from Cedar Lake pond, or deionized water prepared in the laboratory. Germination was evaluated after 24 hours under an inverted microscope. |

**TABLE 7.3 Guidelines for Introductions for Lab Reports and Research Papers**

| | LAB REPORT | SCIENTIFIC RESEARCH PAPER |
|---|---|---|
| **Background** | Some background information | Broad and specific background information and previous research in the area |
| **Unknown/Problem** | Usually not provided | Problems of previous work and unknown factors in the area |
| **Question/Hypothesis of Study** | Indicates purpose of study | Addition made by your research |
| **Experimental Approach** | Approach taken to perform study | Approach taken toward this addition |

In comparison to that of a lab report, the Introduction of a typical journal article/research paper is longer—usually, one to two double-spaced pages or about 250 to 600 words. Longer Introductions, such as those for research papers, contain additional details and background information, as well as a very specific unknown/problem of interest in the respective field to explain the overall purpose of the study (see Table 7.3). Generally, readers expect the parts of the Introduction for research papers to be arranged in a standard structure: a **"funnel,"** starting broadly with background information and then narrowing to what is the problem/unknown, the question, and the experimental/study approach of the paper (see also Zeiger, 2000). Often, the Introduction also gives an overview of the main results and implications of the report or paper. (Note that the Introduction may repeat some parts of the Abstract, which is okay.)

If you are writing a lab report that is more like a full scientific research paper, you will need to provide one to two paragraphs of background information. To do so, you can use lecture notes and descriptions provided in your text or laboratory book. You may also need to do more research using the Internet and library to search recent scientific literature to find other research in this area of study. Summarize that research, stating what the general findings have been and using those findings to describe the current knowledge in the area. For longer lab reports, this background serves to demonstrate that you understand the context for the experiment or study you have completed. The Introduction normally does not include figures or tables.

### ➤ State the research question/hypothesis precisely

The most important element in the introduction is the research question/hypothesis. It should name the variables studied as well as the main features of the study. Note that the question/purpose is usually not written in the form of a question but as an infinitive phrase or as a sentence, using a present tense verb.

**Example 7-6** **Phrasing of question/purpose**

a **To determine** if increasing amounts of $CO_2$ in the atmosphere affect maize seed germination, . . .

b **We examined** the effects of various low temperature durations on the growth rate of the moss *P. patens*.

Following the question in the Introduction of your research paper, briefly indicate your experimental/study approach—usually one sentence, at most two or three sentences. Signal the experimental/study approach so readers can identify it immediately.

**Example 7-7** **Phrasing of the experimental/study approach**

a **We analyzed** restriction enzyme digestion **by** agarose gel electrophoresis.

b The number of eggs per cluster **was characterized by** light microscopy.

## ➤ Signal the elements of the Introduction

Generally, all the parts of the Introduction should be signaled so the reader does not have to guess about the information provided. The signals vary, depending on how the known, unknown, question, and experimental/study approach are phrased. Numerous variations on these signals are possible. Some examples of signals for the required text elements of an Introduction are listed in Table 7.4. You may use these as signals when you compose an Introduction or other sections of your lab report or research paper.

**TABLE 7. 4 Signals of the Introduction**

| BACKGROUND | UNKNOWN | QUESTION, HYPOTHESIS | EXPERIMENTAL/ STUDY APPROACH | RESULTS | IMPLICATION |
|---|---|---|---|---|---|
| X is . . . | . . . is unknown | We hypothesized that . . . | To test this hypothesis, we . . . | We found . . . | . . . consistent with |
| X affects . . . | . . . has not been determined | To determine . . . | For this purpose, we . . . | We determined that . . . | Our findings indicate that . . . |
| X is a component of Y | . . . is unclear | In this study we examined . . . | . . . by/using . . . | Our findings show . . . | . . . may be used to . . . |
| X is observed when Y happens . . . | . . . does not exist | To analyze . . . | To answer this question, we . . . | | . . . is important for . . . |

A good Introduction, in which the background starts very broad but then narrows down quickly to the research topic, is shown in Example 7-8. Note also the signal for the unknown/problem in the last sentence.

**Example 7-8** **Partial Introduction showing good funneling of background**

| | |
|---|---|
| Broad background | The sensory receptors of the auditory system in mammals are the auditory hair cells of the inner ear. Two functionally and structurally different types of mammalian auditory hair cells exist—inner and outer hair cells. While mechanical stimuli are transformed to neural signals in the inner hair cells (Chan and Hudspeth, 2005), outer hair cells do not transmit neural signals to the brain. Instead, when sound enters the inner ear, outer hair cells magnify it mechanically through electromotility, or oscillations at the sound frequency (Brownell et al., 1985). |
| Specific background | The molecular basis of this mechanism is thought to be the motor protein prestin, which is embedded in the lateral membrane of the outer hair cells. Mammalian prestin is an 80-kDa, 744-amino-acid membrane protein whose function appears to depend on chloride channel signaling (Santos-Sacchi et al., 2006). |
| Unknown/ Problem (signal in italics) | Although prestin has been researched intensively, its molecular function *has not been fully established.* |

A complete Introduction, containing all elements, is shown in Example 7–9. Note how this Introduction funnels from broad to specific background to the unknown, the question, and the experimental/study approach. The latter two elements are stated in the last paragraph of the Introduction, which represents an important power position of this section. Here, too, note the signals used to identify the different elements of the Introduction.

**Example 7-9** **Complete Introduction containing all elements**

| | |
|---|---|
| Broad to specific background (signal in italics) | The species-specific attachment of sperm to sea urchin eggs *provides a model system* for the biocheminal study of intercellular adhesion. The protein bindin is located in the acrosome granule of sea urchin sperm and is known to be the sperm component of the sperm–egg bond (Vacquier and Moy, 1977). Bindin proteins isolated from two different species of sea urchins show similar yet distinct amino acid compositions and sequence (Bellet et al., 1977). |
| Problem (signal in italics) | Further study of *S. purpuratus* species-specific bindin and its interaction with eggs requires large amounts of the protein. As isolation from sea urchins is inefficient, *a new method of production had to be found.* |

| | |
|---|---|
| More background | The use of bacteria for protein expression has been a powerful tool of genetic engineering. Not only does this use of bacteria allow production of proteins in biochemically useful quantities, but also the systematic alteration of the protein structure by site-directed mutagenesis of the cDNA and subsequent study of the effect of such alterations on the function of the protein *in vitro*. |
| Question/ purpose and experimental approach (signals in italics) | *In this study, we expressed* the sea urchin protein bindin in *E. coli in order to* isolate it in large amounts for functional studies. The expression vector developed to accomplish this could also be used for future site-directed mutagenesis experiments. |

## 7.8 MATERIALS AND METHODS

### ➤ Provide sufficient details and references in the Materials and Methods section

The Materials and Methods section describes the experimental/study approach used to arrive at your conclusions. It does so by relating important details such as materials and methodological approaches in a narrative. It usually does not include figures or tables unless absolutely necessary to relay study set-ups. The Materials and Methods section should not be written in the form of a recipe (1. Obtain oxygen meter; 2. Place transducer in sample), nor should it be a mere laundry list of materials (1 bucket, 2 feet of string, etc.) Instead, you need to tell a story, including why you did something, how you did it, what materials you used, and so forth.

The Materials and Methods section should contain sufficient detail to allow a trained scientist to evaluate or repeat your work. You should write this section with care so that your experimental/study approach does not appear faulty, incomplete, or unprofessional.

The Materials and Methods section should cover:

- Materials (drugs, culture media, buffers, gases, or apparatus used)
- Subjects (patients, materials, animals, microorganisms, plants)
- Design (includes independent and dependent variables, experimental/study and control groups)
- Procedures (what, how, and why you did something)

Define materials and methods as precisely as you can. You need to provide sufficient details and exact technical specifications such as temperature, pH, total volume, time, and quantities to ensure that scientists can repeat your work. Do not forget to include your control experiments and groups.

Following are a few examples that do not provide sufficient detail:

**Example 7-10** **Providing insufficient detail**

**a** To characterize the native flora, orchids were collected.

**b** The number of eggs per cluster was characterized by light microscopy.

**c** All samples were centrifuged.

**Revised Example 7-10** **Providing sufficient detail**

**a** To characterize the native flora, orchids were collected **in the Everglades during January 2014.**

**b** The number of eggs per cluster was **counted** under light microscopy **for all the different individual male and female fertilization combinations using gamete concentrations of 1000 per micro liter.**

**c** All samples were centrifuged **at 5,000 x g for 30 min at 25°C.**

Be sure not to go overboard providing too much detail, such as tube or pipette size or color, or the manufacturer's name unless this is relevant to your study.

**Example 7-11** **Unnecessary information in Materials and Methods**

Cells were scraped out of the wells and resuspended in a 1.5 ml Eppendorf tube.

**Revised Example 7-11** Cells were scraped out of the wells and resuspended **in 100 µl of sterile saline.**

## ➤ Divide the Materials and Methods section into subsections

You can organize your Materials and Methods section by separating groups of actions into subsections, each with its own subheading. Although use of subsections is optional, it usually simplifies and clarifies the presentation for the reader. The sequence of events within these subsections is then written in chronological order or from most to least important.

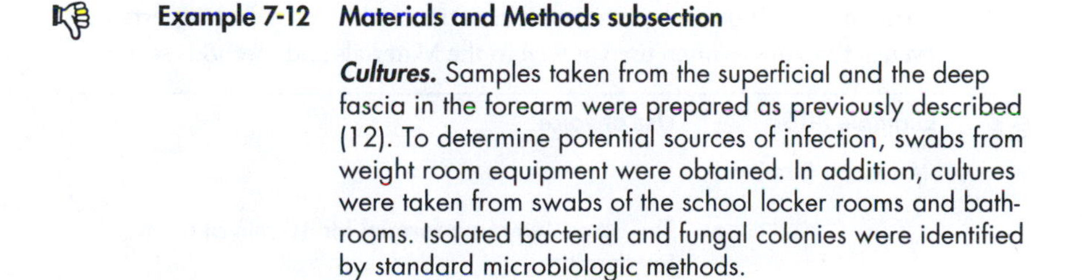

**Example 7-12** **Materials and Methods subsection**

***Cultures.*** Samples taken from the superficial and the deep fascia in the forearm were prepared as previously described (12). To determine potential sources of infection, swabs from weight room equipment were obtained. In addition, cultures were taken from swabs of the school locker rooms and bathrooms. Isolated bacterial and fungal colonies were identified by standard microbiologic methods.

### ➤ Provide literature references if needed

If your methods have not been reported previously, you must provide all of the necessary detail. If, however, methods have been described previously, such as in a lab manual, you may provide only that literature reference. If you modified a previously published method, provide the literature reference and give a detailed description of your modifications.

---

**Example 7-13** **Referring to a previously described method**

Plasmids were isolated as described in the Bio 112 lab manual.

---

**Example 7-14** **Referring to a previously described method with modifications**

Plasmids were isolated as described in the Bio 112 lab manual with minor modifications. Instead of dissolving DNA pellets in sterile water, pellets were dissolved in buffer A.

---

### ➤ State the purpose or reason for an experiment or observational study part if needed

It is important to ensure that the reader will understand why each procedure was performed and how each procedure is linked to the central question of the paper. Therefore, you should state the purpose or give a reason for any procedure whose function or relation to the question of the paper is not clear.

---

**Example 7-15** **Statement of purpose**

**a.** **To purify X,** the mixture was run over a Y column.

**b.** **To assess the frequency of feedings,** we recorded the number of parental visits over a period of 4 weeks.

---

### ➤ Use the appropriate voice and tense

The Materials and Methods section is the one section in a research paper where often passive voice is preferred over active voice. The reason is twofold: It lets you emphasize materials or methods as the topic of your sentences, and readers do not need to know who performed the action. In less formal lab reports, however, active voice is often used as well in the Materials and Methods section.

---

**Example 7-16** **Use of voice**

**Passive**

**a.** The assays **were performed** for 10 min at room temperature.

**b.** Young **were observed** every other day for two months.

**Active**

a. We **performed** the assays for 10 min at room temperature.

b. We **observed** the young every other day for two months.

---

In the Materials and Methods section, the general rule about verb tense applies (see Chapter 2). When reporting completed actions, use past tense. However, use present tense for statements of general validity and for those whose information is still true or if you are referring to figures and tables.

---

**Example 7-17** **Use of tense**

a. Cicadas **were collected** in August.

b. Cicadas **are hunted** by praying mantises.

c. Criteria used in selecting subjects **are listed** in Table 2.

---

## 7.9 RESULTS

### ➤ Report your main findings in the first paragraph of the Results section

The Results section presents the results of your study and points the reader to the data, which support the results and are shown mainly in the figures and tables. Do not forget to include control results and, if needed, explain the purpose of an experiment or study portion shortly.

Start the Results section by presenting your main findings in the first paragraph. Your main findings are the findings used in providing the overall answer/conclusion of the paper. You may also start the first paragraph with a brief overview of your general observations and then move on to the main findings. In the latter case, do not devote more than a few sentences to any overview, and ensure that your main findings still appear in the first paragraph because it is a power position.

---

**Example 7-18** **Results—first paragraph**

| | |
|---|---|
| Overall background/purpose and experimental approach | During fertilization experiments performed to determine the fertilizability of the sea urchin species *S. purpuratus*, |
| Main overall results (signal in italics) | *we discovered* that as many as one third of the sea urchins examined appeared to be infertile. Furthermore, 0.5% of the urchins were hermaphrodites, that is, they contained eggs and sperm. Substantial variations in fertilization were observed for approximately 30% of the individual urchins (Fig. 2). |

**TABLE 7.5 Guidelines for the Results Section of Lab Reports and Research Papers**

| | LAB REPORT | SCIENTIFIC RESEARCH PAPER |
|---|---|---|
| **First paragraph** | Present main results | Provide an overview of main results |
| **Subsequent paragraphs** | Elements<br>• Purpose or background of experiment/study portion if needed<br>• Experimental/study approach<br>• Results | Elements<br>• Purpose or background of experiment/study portion if needed<br>• Experimental/study approach<br>• Results<br>• Interpretation of results |

### ➤ Structure each experimental part in subsequent paragraphs by stating:

- Purpose or background of experiment (1/2 to 1 sentence)
- Experimental approach (1/2 to 1 sentence)
- Results (1 to multiple sentence(s))
- Interpretation of results (for research papers) (1/2 to 1 sentence)

In subsequent paragraphs, present your specific observations. Information in these paragraphs needs to be organized. For simple lab reports, each paragraph that describes results of a specific individual experiment or observation should also contain the following essential components:

- Purpose or background of experiment if needed
- Experimental/study approach
- Results

Start your segments or paragraphs by providing a topic sentence. This topic sentence usually indicates the purpose of the experiment or observation performed. It may also provide context in form of background information. Do not assume that the reader remembers anything from the Introduction or Materials and Methods sections. The purpose may be written in the form of a transitional phrase or clause, for example, and is followed by an equally short statement of your experimental/study approach (1/2 to one sentence is sufficient). This short description, too, is to provide context for the reader, who may not even have read the Materials and Methods section. Follow the approach immediately with your results. Place important or general results first and less important details later in the paragraph.

### ➤ Include a brief interpretation of the findings

For longer, more in-depth lab reports and for research papers, a general interpretation of the results should be added to each paragraph or portion to make the results meaningful for the reader. Note that this last sentence in a results paragraph should only state what the results mean without discussing them further—a detailed discussion will follow in the Discussion section.

In longer lab reports or research papers, you may even consider dividing your Results section into different subsections to make navigating through the section easier for the reader. Ensure, however, that you report only results that are pertinent to the question posed in the Introduction and to the experiments and observations described in the Materials and Methods. Exclude preliminary results and results that are not relevant. Also incorporate results whether or not they support your hypothesis, and explain any contradicting results if necessary.

| **Example 7-19** | **Well-formed Results paragraph (signals in italics)** |
|---|---|
| Short background | *Nocardia have* a unique cell wall and membrane structure, which may be permeable to growth inhibitors. |
| Short purpose | *We therefore tested Nocardia's* permeability to growth inhibitors |
| Short experimental approach | *by* adding different inhibitors to the bacterial medium before incubation. |
| Results | *We found that* two out of the five molecules tested inhibited growth of the bacterial cultures. NB22 demonstrated 100% growth inhibition at 15 μg/ml, whereas NB20 inhibited growth only weakly at the same concentration. Bacterial growth on the negative control plate was not affected. |
| Interpretation of results | *These results indicate that* the cell wall and membrane structure of *Nocardia* is fully permeable to the growth inhibitors NB22 and only partially permeable to NB20. |

Regardless of whether you are composing a Results section for a lab report or one for publication in a journal, throughout the Results section, emphasize the data and their meaning. Subordinate control results and methods.

### ➤ Distinguish between data and results

To present your results to the reader clearly, you need to distinguish between data and results. Data are values derived from scientific experiments or observations (concentrations, absorbance, mean, percent increase). Results interpret data; that is, they describe trends and patterns in the data, which in turn are then further interpreted and discussed in the Discussion section:

**Example 7-20** Absorbance **increased** 55% when samples were incubated at 25°C instead of at 15°C.

If data are provided without explanation, findings are not very meaningful to readers.

| | | |
|---|---|---|
| **Example 7-21** | **Presenting data without interpretation** | |
| | Among the 785 obese mice, we found 522 males and 163 females. | |

To explain the data for the reader, the author needs to first present the interpretation and then the data that support it.

| | |
|---|---|
| **Revised Example 7-21** | We found that **3.8 times as many** male mice (79.2%) than female mice (20.8%) were obese. |

When you state your interpretations/results, ensure that you make reference to your data in figures or tables by referencing the figure or table number in parentheses where needed.

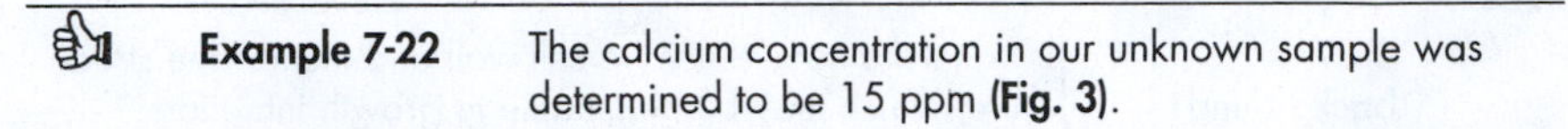

| | |
|---|---|
| **Example 7-22** | The calcium concentration in our unknown sample was determined to be 15 ppm **(Fig. 3)**. |

Although most data should be presented in figures and tables, your main findings should be stated in the text as well along with their explanation.

### ➤ Report results in past tense

Results are usually reported in past tense because they are events and observations that occurred in the past.

| | |
|---|---|
| **Example 7-23** | **Use of tense** |
| | (a) Compound X **inhibited** conversion of A to B. |
| | (b) All hawkmoths **started** feeding from the *A. palmeri* flowers only after nectar from the *D. wrightii* flowers **was depleted (Riffell, 2008)**. |

### ➤ Signal the elements of the Results section

As in the Abstract and Introduction, the different elements in the Result section should be signaled so that readers cannot miss them. To emphasize different elements, use signals such as those listed in Table 7.6.

**TABLE 7.6 Signals for the Results**

| PURPOSE/QUESTION | EXPERIMENTAL/STUDY APPROACH | RESULTS | INTERPRETATION OF RESULTS |
|---|---|---|---|
| To determine . . . | . . . we did . . . | We found . . . | , indicating that . . . |
| To establish if . . . | X was subjected to . . . | We observed . . . | , consistent with . . . |
| Z was tested . . . | . . . by/using . . . | We detected . . . | , which indicates that . . . |
| For the purpose of XYZ . . . | ABC was performed . . . | Our results indicate that . . . | This observation indicates that . . . |
| | Experiment X showed . . . | | A is specific for . . . |

## 7.10 DISCUSSION

### ➤ Organize the Discussion into first, middle, and final/concluding paragraph(s)

| | |
|---|---|
| First paragraph: | Interpretation/answer based on key findings |
| | Supporting evidence |
| Middle paragraphs: | Secondary results |
| | Limitations of your study |
| | Unexpected findings |
| | For research papers also: |
| | Comparisons/contrasts to previous studies |
| | Hypotheses or Models |
| Final/concluding paragraph: | Summary |
| | Significance/implication |

The Discussion is usually the hardest section to define and to write. Even though the data may be valid and interesting, the interpretation or presentation of it in the Discussion may obscure it. Therefore, good style and clear, logical presentation are especially important here.

In the first paragraph, tell your readers what your key findings were and what they mean. In subsequent paragraphs, explain how your findings fit into what is known in the field. In the last paragraph, summarize and generalize why the contribution of your study is important overall, in your field, outside your field, and/or for society.

### ➤ Answer the research question in the first paragraph

In the Discussion, provide the answer to the research question. To do so, interpret your key findings and draw conclusions based on these findings—in other words, answer the question(s) asked in the Introduction. The Discussion should also explain how you arrived at your conclusions.

Because the interpretation of your key findings is the most important statement in the paper, it should appear in the most prominent position: the first paragraph of the Discussion. The interpretation of your key findings should match the question/purpose for the study stated in the Introduction and answer what the Introduction asked. The interpretation of your key findings should also be repeated in the other power position of this section: the last paragraph.

Providing more background information, summarizing your findings, or reporting on limitations and minor results are not good ways to start the Discussion. Instead, begin by directly stating the answer based on your findings in the opening sentence of the Discussion. If you feel that this beginning is too abrupt, you can restate the purpose of the study or provide a brief context before stating the answer. Any statements placed before your answer should not exceed more than a few sentences.

---

**Example 7-24** **First paragraph of Discussion**

**Question/Purpose:**

Our goal was to determine **what part of the *bindin* polypeptide is responsible for the species-specific egg agglutination activities of the protein.**

**Answer/Interpretation of Key Findings:**

| | |
|---|---|
| Answer to question (signal in italics) | *Our results suggest that* **the part of bindin responsible for species-specific egg agglutination lies in the region of residues 75–121**. *We showed that* **residues 18–74 and 122–236 can be deleted without loss of egg agglutination activity**. |
| Supporting evidence (signal in italics) | All of the biologically active binding deletion analogs were found to be **species-specific** by their ability to **agglutinate** exclusively *S. purpuratus* eggs. Deletion analogs that had any residues of region 75–121 deleted exhibited no significant activity above the bacterial control protein. |

(*With permission from Elsevier*)

---

In Example 7-24, the interpretation of the key findings matches the purpose of the study. Note that the same key terms ("bindin," "species-specific," "egg agglutination") appear in the question/purpose as well as in the interpretation or answer to the question. This answer is immediately supported by key findings of the study.

### ➤ Summarize and generalize your results in subsequent paragraphs

In the Discussion, you should include explanations for any results that do not support the answers. In addition, you may discuss any possible errors or limitations in your methods, give explanations of unexpected findings, and indicate what the next steps might be. Do not refer to every detail of your work again. Instead, in your Discussion, summarize and generalize.

After stating and supporting your answer, mention other findings that were important. Tell your readers what you think your results mean and how strongly you believe in them. Organize these findings according to the science or from most to least important. To ensure that your Discussion is organized rather than rambling, focus the story on the question/purpose of the paper that was stated in the Introduction. Treat your secondary results as you did your main findings: Summarize and generalize them rather than simply repeating what you found.

In these subsequent paragraphs, you may also mention any limitations of your study or unexpected findings and may present any new hypothesis or model based on your findings. See, for instance, Example 7-25, which explains limitations of a given study and delineates what the outcomes would be under a different set of assumptions than used for the given study. If useful, include figures to illustrate complex models in the Discussion.

| **Example 7-25** | **Explaining limitations in the Discussion** |
|---|---|
| Limitation (signals in italics) | In our modeling of Aβ assembly, we assumed that Aβ monomers are not present in drusen. However, *it is possible that* Aβ monomers, once polymerized into amyloid fibrils, *may accumulate* in drusen (40). Such accumulations *would result* in a lower number of monomers used for calculations in our model than are actually present, and thus a higher risk for the disease than determined based on our assumption. |

For in-depth lab reports or research papers, compare and contrast your findings with those of previously published papers, but avoid the temptation to discuss every previous study in your subject area (see Table 7.7 for the difference in the Discussion section of lab reports and research papers). Stick to the most relevant and most important studies. Explain any disagreements objectively, and credit and confirm the work of others by referencing it. Give pro and con arguments for your conclusion. Only if you mention both impartially will you sound convincing to the reader. You may also state theoretical implications or practical applications. Know that most of the time it is wise to present your opinion carefully rather than too strongly.

**TABLE 7.7 Guidelines for the Discussion Section of Lab Reports and Research Papers**

| | LAB REPORT | SCIENTIFIC RESEARCH PAPER |
|---|---|---|
| **First paragraph** | Present main results and their interpretation | Present main results and their interpretation |
| **Subsequent paragraphs** | Provide:<br>• Secondary results and their interpretation<br>• Limitations of your study<br>• Unexpected findings | Provide:<br>• Secondary results and their interpretation<br>• Comparisons/contrasts to previous studies<br>May also provide:<br>• Limitations of your study<br>• Unexpected findings<br>• Hypotheses or models |
| **Conclusion** | Summarize main findings and indicate their importance | Summarize main findings and indicate their importance |

| | | |
|---|---|---|
| **Example 7-26** | **Formulating hypotheses** (note signals for results, unknown, and hypothesis in italics) |
| Hypothesis | *We found* that the substrate $^{3}$H-[9R]iP moves into the cells where it does not accumulate to concentrations higher than in the medium. However, the mechanism of $^{3}$H-[9R] iP uptake *is unclear*. Because no extracellular activities for the deribolization of $^{3}$H-[9R]iP could be detected, *we hypothesize* that it is metabolized intracellularly to $^{3}$H-iP and that the bidirectional transport of iP is based on passive diffusion. |

### ➤ Provide a conclusion

At the end of the Discussion, you should provide some closure by writing a one-paragraph concluding summary. Readers typically expect to see two things in the summary of a scientific paper: an analysis of the most important results and the significance of the work. The analysis of the most important results is typically provided by the interpretation of your key findings—that is, the answer. Here, too, the answer should match the question/purpose you posed in the Introduction and the answer presented in the first paragraph of the Discussion. Do not bring in new evidence for the summary. Rather, complete the "big picture" by restating your answer—that is, the interpretation of the key findings.

| | |
|---|---|
| **Example 7-26** | **Conclusion** (note signals in italics) |
| Conclusion | *In summary*, fertilization among sea urchins appears intraspecies specific due to surface components of both gametes. *It is conceivable that* these surface components contain multiple adhesive elements. Based on our hypothesis, mutations in these elements *may result in* reproductive isolation and speciation. |

### ➤ Signal the different elements in the Discussion

Signal the different elements of the Discussion so that readers recognize immediately what they are reading about. Possible signals for various elements are listed in Table 7.8.

**TABLE 7.8 Signals for the Discussion**

| ANSWER | KEY FINDINGS | SUMMARY | SIGNIFICANCE |
|---|---|---|---|
| In this study, we have shown that . . . | In our experiments . . . can be attributed to . . . | In summary, . . .<br>In conclusion, . . .<br>Finally, . . . | Our findings can/will serve to . . .<br>. . . can be used . . . |
| In this study, we found that . . . | We determined X by . . . | Taken together, . . . | We recommend that X is . . .<br>Y should be used for . . . is probably . . . |
| Our study shows that . . . | We found that . . . | To summarize our results, . . . | Y indicates that X might . . . |
| Our findings demonstrate that . . . | Our data shows that . . . | We conclude that . . . [overall question] | These findings imply that X may . . . |
| This paper describes . . . | . . . has been demonstrated by . . . | Overall, . . . | Here we propose that . . .<br>We hypothesize that . . . |

## 7.11 REFERENCES

### ➤ Cite your sources

In all research papers and lab reports, your interpretation of the available sources must be backed up with evidence. Therefore, cite primary and secondary sources where needed and include a reference list. The type of information you choose should relate directly to the article's focus. See Chapter 4 for more information on reference selection and citation.

## 7.12 ACKNOWLEDGMENTS

### ➤ Acknowledge organizations and individuals if needed

Lab reports usually do not contain an Acknowledgments section, but research papers do. In this section, acknowledge any organizations or individuals who provided grants, materials, and financial or technical assistance as well as those who contributed ideas, information, and advice to your work. You do not need to acknowledge anyone who just did their day-to-day work. People named in the Acknowledgments should give their permission to be named and should approve of the wording of your acknowledgment.

For the order of acknowledgments, start by listing intellectual contributions, then move on to technical support, provision of materials, helpful discussions, and revisions and preparations of the manuscript. Last, list any funds, grants, fellowships, or financial contributions.

**Example 7-27** We thank Dr. J. Holzheimer for his technical advice on crystallization assays. We are also grateful to Drs. Thomas Hugh and Fred Grant for their critical review of the manuscript. This study was supported by grant XXX from the National Institutes of Health.

## 7.13 CHECKLIST

Use the following checklist to ensure that you have addressed all important elements for a lab report or research paper. In your revisions and in editing papers, work your way backward from paragraph location and structure to word choice and spelling.

### Abstract

- ☐ Is the question/purpose stated precisely?
- ☐ Is the approach stated?
- ☐ Are the results indicated?
- ☐ Is the answer/conclusion provided?
- ☐ Are all elements signaled?
- ☐ Is the length within the required limits?
- ☐ Is the significance of the work apparent?

### Introduction

- ☐ 1. Are all the components there?
  - ☐ a. Background
  - ☐ b. Unknown
  - ☐ c. Question/purpose
  - ☐ d. Approach
  - ☐ e. Results/conclusion
  - ☐ f. Significance
- ☐ 2. Is the research question stated precisely? (Is it in present tense?)
- ☐ 3. Do all the components logically follow each other? (Is the unknown what one would expect to hear after reading about what is known? Is the research question really the question one would anticipate based on the unknown? Does the answer really answer the research question?)
- ☐ 4. Is all background information directly relevant to your research question? Did you list only the most pertinent literature and not review the topic?
- ☐ 5. Have all elements been signaled clearly?
- ☐ 6. Are references placed correctly and where needed? (Chapter 4)
- ☐ 7. Is the Introduction cohesive and coherent? (Chapter 3)

### Materials and Methods

- ☐ 1. Do the listed materials and methods describe all procedures and approaches done to obtain the results presented?
- ☐ 2. Are sufficient details and/or references provided?
- ☐ 3. Are protocols logically grouped and organized?
- ☐ 4. Are topics signaled and linked?
- ☐ 5. Did you pay attention to voice (mainly passive)?

- ☐ 6. Did you ensure that major results are not stated in the Materials and Methods section?
- ☐ 7. Is the purpose stated for any procedure whose function is not clear?

**Results**

- ☐ 1. Did you report all main findings as well as other important findings?
- ☐ 2. Are your most important results and their interpretation provided in the beginning of the Results section?
- ☐ 3. Does each Results segment or paragraph contain all components (purpose of study, approach, results, and their interpretation)?
  - ☐ a. Is the purpose of each study apparent?
  - ☐ b. Is the approach provided?
  - ☐ c. Are results interpreted?
- ☐ 4. Are all components (purpose of study, approach, results, and their interpretation) signaled?
- ☐ 5. Is the reader pointed to figures and tables?
- ☐ 6. Are control results included?

**Discussion**

- ☐ 1. Did you interpret the key findings?
- ☐ 2. Is the interpretation/answer to the research question in the first paragraph?
- ☐ 3. Is the answer followed by supporting evidence?
- ☐ 4. Is a summary paragraph placed at the end of the Discussion?
- ☐ 5. Is the significance of the work apparent?
- ☐ 6. Did you compare and contrast your findings with those of other published results?
- ☐ 7. Did you explain any discrepancies, unexpected findings, and limitations?
- ☐ 8. Did you provide generalizations where possible?
- ☐ 9. Did you avoid restating or summarizing the results?
- ☐ 10. Are all elements signaled?

## SUMMARY

### WRITING A LAB REPORT OR RESEARCH PAPER

1. Understand the "big picture."
2. Understand what is involved.
3. Obtain instructions and follow them.
4. Collect, organize, and study your references.
5. Follow the IMRAD format.
6. Focus on the overall question or hypothesis of the study.

7. Make the title succinct, clear, and complete.
8. Avoid nonstandard abbreviations in the title.
9. Format the title page correctly.
10. Include the following elements in the Abstract: question/purpose, experimental/study approach, results, and conclusion.
11. Do not include general overview sentences, nonstandard abbreviations, references, figures, or tables in the Abstract.
12. Signal the elements of the Abstract.
13. Follow a funnel structure for the Introduction and include: background (unknown for research papers only), question/hypothesis, experimental/study approach, optional: results and significance.
14. State the research question/hypothesis precisely.
15. Signal the elements of the Introduction.
16. Provide sufficient details and references in the Materials and Methods section.
17. Divide the Materials and Methods section into subsections.
18. Provide literature references if needed.
19. State the purpose or reason for an experiment or observational study part if needed.
20. Use the appropriate voice and tense in the Materials and Methods section.
21. Report your main findings in the first paragraph of the Results section.
22. Structure each experimental part in subsequent paragraphs by stating: purpose or background of study, experimental/study approach, results, interpretation of results (for research papers).
23. Include a brief interpretation of the findings.
24. Distinguish between data and results.
25. Report results in past tense.
26. Signal the elements of a Results section.
27. Organize the Discussion into first, middle and final/concluding paragraphs.
28. Answer the research question in the first paragraph of the Discussion.
29. Summarize and generalize your results in subsequent paragraphs.
30. Provide a conclusion at the end of the Discussion.
31. Signal the different elements in the Discussion.
32. Cite your sources.
33. Acknowledge organizations and individuals if needed.

## PROBLEMS

### PROBLEM 7-1 Abstract

**In the following Abstract, identify all essential components and their signals if provided:**

To remove cellular metabolic wastes from the body, mammals have an excretory system. To gain more insight into such a system, we dissected a kidney of a fetal pig. For this purpose, we cut from the lateral margin, leaving the ureter intact. We found that the kidney consists of three different internal regions: the outer cortex, the middle medulla (containing the renal columns and pyramids), and the inner renal pelvis. The kidney's nephrons filter water, nitrogenous wastes, and other materials from the blood, which enters the kidney through the renal artery above the renal vein. Water and waste material are collected and flow as urine through the collecting ducts into the renal pelvis. Ultimately, urine is then passed through the ureter, which empties into the urinary bladder. Filtered blood flows back to the body through the renal vein. The study of the kidney shows how mammals eliminate their fluid waste.

### PROBLEM 7-2 Introduction

**Why does this Introduction seem incomplete? Identify the known, unknown, question/purpose, and approach. Are these elements clearly identifiable? Why or why not?**

The carotenoid astaxanthin is a red pigment that occurs in specific algae, fish, crustaceans, and in some bird plumages (McGraw and Hardy, 2006). Astaxanthin is an antioxidant and commonly is used as a natural food supplement, food color, and anticancer agent among other disease preventative measures.

Like many carotenoids, astaxanthin is a colorful, fat/oil-soluble pigment, providing a reddish and pink coloration (2). Whereas in certain bird species all adult members display carotenoid-containing feathers rich in color, many gulls and terns, which normally have white feathers, display an abnormal pink tinge (or flush) in various degrees across their populations (McGraw and Hardy, 2006). It has been suggested that this pink tinge arises during feather growth when these birds ingest abnormally high quantities of astaxanthin, which often occurs close to salmon farms (McGraw and Hardy, 2006). However, the exact relationship between astaxanthin and plumage is not fully understood. Here, we examine this relationship in more detail and discuss its implication.

### PROBLEM 7-3 Results

**Assess and revise the following partial Result section. Ensure that all the parts of a paragraph for the Results section are provided and any unnecessary parts are omitted.**

To evaluate inhibitory effects of the isolated molecules, 10 mM stock solutions of all isolates were prepared in buffer A. The buffer was previously optimized for the inhibition assay and contained 25 mM Tris-HCl (pH 7.5), 5 mM β-glycerophosphate, 2 mM dithiothreitol (dTT), 0.1 mM $na_3Vo_4$, 10 mM $MgCl_2$, and 250 μM ATP. For the assay, 0.1 mM of the isolated molecule were used in 200 μl total volume. The reaction mixture was vortexed for 30 sec before incubation at 25°C for 1 hour. Then, kinase was added to 100 μM. Reaction products were analyzed by 12% PAGE and Western blot analysis.

Six out of 15 isolated molecules inhibited the kinase reaction at 10 μM markedly, and two of the molecules, A3 and A7, exhibited more than 50% inhibition of the enzyme activity. A 10-fold dilution series of these latter samples was prepared to determine the minimal inhibitory concentration. Molecule A3 exhibited 30% and molecule A7 45% inhibition of the kinase reaction at 1 μM. Buffer A was found not to interfere with the enzyme.

**PROBLEM 7-4 Results**

**Assess the following partial Results section. Then:**

1. **Identify**
   - **the purpose or background of the study**
   - **the approach**
   - **the results**
   - **the interpretation of the results**
2. **Are all the parts of a paragraph for the Results section provided? Please explain.**

We found that the H384A mutant reduced the $k_{cat}$ value more than 3-fold. The apparent $K_m$ values were increased 7-fold for Fru 6-P and 3.5-fold for PPi. The increase of the $K_m$ values and the reduction of the $k_{cat}$ value of the H384A mutant suggest that the imidazole group of His384 is important for the binding stability as well as for catalytic efficiency of Fru 6-P and PPi substrates.

**PROBLEM 7-5 Discussion**

**Consider the two different opening paragraphs of a Discussion about a preventive measure against malaria. Which one is a better first paragraph for a Discussion and why?**

**Version A**

We trapped and counted the number of mosquitoes within the urban environment of the city of Kumasi using conventional carbon dioxide traps. Nearly 70% more adult *A. gambiae* were caught in communities near moist urban agricultural establishments than in rural locations or in non-irrigated urban settlements. When we evaluated malaria episode reports from people living in various parts of the city, we found that 18% of malaria cases in all seasons were reported by those near urban agricultural sites, whereas only 2% of the control groups reported incidences of malaria per year.

**Version B**

The results of this study show that open-space irrigated vegetable fields in cities can provide suitable breeding sites for *A. gambiae*. This is reflected in higher numbers of adult *A. gambiae* in settlements in the vicinity of irrigated urban agricultural sites compared to control areas without irrigated urban agriculture. Moreover, people living in the vicinity of urban agricultural areas reported more malaria episodes than the control group in the rainy as well as dry seasons. Apparently, the informal irrigation sites of the urban agricultural locations create rural spots within the city of Kumasi in terms of potential *Anopheles* spp breeding sites.

*(With permission from Elsevier)*

**PROBLEM 7-6 Discussion**

**Consider the two different concluding paragraphs of a Discussion about desert frogs. Which one is a better conclusion for a Discussion and why?**

**Version A**

In conclusion, this study shows that desert frogs can avoid death by desiccation by maintaining a high body water content and water storage in their urinary bladder and by rapid hydration when water is available. These measures may be employed in combination with behavioral adaptations such as burrowing and change in pigmentation to minimize stresses tending to dehydrate the animals.

**Version B**

A limitation of this study was the small number of animals, a single species of frogs, and the location of the study area, which took place in only one oasis in the Mohave Desert. Future studies should be extended to other species, a larger number of animals, and a greater diversity of locations.

CHAPTER 8

# Revising and Editing

THIS CHAPTER EXPLAINS:

- How to revise your first draft
- How to revise any subsequent drafts
- How to edit someone else's manuscript
- The submission and review process

## 8.1 GENERAL ADVICE

Revision is the key to successful writing. Do not expect to finish writing and revising your document in one go. Most documents will need several revisions, and putting off a report until the night before a due date will not allow enough time to carefully revise a paper. Generally, the more important the document, the more it should be revised. Professional documents, in fact, often have more than ten revisions, spread over several weeks or months, before they are submitted. Recognize that good writing takes time, patience, and a lot of hard work.

## 8.2 REVISING THE FIRST DRAFT

### ➤ Check the first draft for content and content location

When you revise your first draft, check it first for content and organization. Make sure that all the essential points have been included. Everything you say should contribute in some way to the overall purpose and findings (see also Chapter 7), and no steps should have been left out. Any irrelevant points need to be removed, and any missing evidence should be included.

In revising, you essentially work your way from the larger structures of the paper down to the smaller structural elements. Therefore, check the content and organization of the individual sections of the paper first

(Introduction, Materials and Methods, Results, Discussion), then paragraphs, and then sentences. All the parts, paragraphs, and sentences must be in the right order before you revise further, such as for style.

The overall structure of your paper should conform to the following outline:

| | | |
|---|---|---|
| **Title:** | **3–4 important key terms** | |
| **Abstract:** | **Content: question/purpose, experimental/study approach, results, interpretation/answer, significance** | |
| **Introduction:** | Organization: funnel shape (from broad to specific background information of what is **known**, what remains **unknown**, the **question** you are addressing, the **experimental/study approach**) | |
| | **First paragraphs: Background** | |
| | **Second to last paragraph: Unknown** | |
| | **Last paragraph:** | **Question/purpose and experimental/study approach. Optional are main results and significance.** |
| **Materials and Methods:** | | Organize chronologically, most to least important, or by subsections |
| **Results:** | **1. Paragraph(s):** | **Overview of most important/ interesting result(s)** |
| | Middle paragraphs: | Describe results in detail. Organize chronologically or most to least important—every result segment should contain the purpose of the study, experimental/study approach, results, and their interpretation |
| | Last paragraph: | State interesting result(s) or summarize main findings if Results section is lengthy |
| **Discussion:** | Organization: pyramid shape (from specific to more general; interpretation of findings, compare and contrast, models, conclusion, significance) | |
| | **1. Paragraph** | **Interpret most important results/answer to the question of the paper; support and defend interpretation** |
| | Middle paragraphs: | Chain of topics, compare and contrast findings, list limitations, etc. |
| | **Last Paragraph:** | **Conclusion: Summarize main findings and significance (future directions)** |

### ➤ Pay special attention to key power positions

Pay particular attention to the content and location of power positions throughout your revisions. These key power positions are written in bold in the prior overall outline. They indicate the most crucial structural locations of a paper or report.

Use the checklist provided for each section of a paper at the end of Chapter 7 to double check that you have included all relevant components in each section. It may help to mark and label important components of each section on a print version of the manuscript to check their completeness. Identifying each essential component will make it apparent to you if there is anything missing or not clearly signaled. Above all, the purpose of your study and the interpretation of your results have to make sense together.

In addition to the overall order listed above, you should consider a secondary order of your topics within each section of your paper. For example, if you are addressing multiple questions in a particular order, keep that order throughout the paper. List your analyses, results, and discussion components in the same order that you introduced your questions. Such order makes it easier for the readers to follow your thoughts.

### ➤ Check logical organization and flow of sections and subsections

When you are happy with the structural organization and content of power positions, revise each section for logical organization and flow. Ensure that headings refer to the text they describe. Look at how the ideas are distributed among the paragraphs, and make sure that your arguments are logical. Is it clear how and why the evidence presented supports the interpretation of your findings? Is it clear why a particular experimental/study approach or technique is appropriate? Have the main concepts been clearly and logically connected?

To check for logical flow, verify that you have a chain of topic sentences running throughout the paper. When read by themselves, the topic sentences should be sufficient to provide a rough outline of the paper.

It may also help to make a reverse outline of your manuscript by going through it paragraph by paragraph. Check that this reverse outline is logically organized.

### ➤ Revise for style only after you are satisfied with the content and organization

Once you are satisfied with the content and organization of the first draft, revise it stylistically. You will probably see a lot to change. Here, too, start by working your way from the bigger structures toward the smaller ones. Refer to the summaries of Chapters 2 and 3 as the checklists for reference of style and composition.

### ➤ Pay particular attention to key terms and transitions

For good flow between sections and between paragraphs, ensure that the transitions between paragraphs and sections are smooth and that they tie the pieces together. Pay particular attention to key terms and transitions within paragraphs. Add transition phrases and clauses to create the overview of the story (see also Chapter 3, Section 3.2). Then, use the basic writing rules discussed in this book to check for paragraph structure, sentence structure, and word choice. Inch through your manuscript sentence by sentence, word by word. Consider word location (see Chapter 3, Section 3.2). Check whether you have paid attention to either jumping word location or a consistent point of view.

### ➤ Condense where possible

Look for all possible ways to condense your paper: Omit needless details, redundant words, vague words, and unnecessary paragraphs. If the same concept can be communicated using fewer words, then edit it to the shorter version (see also Chapter 3, Section 3.3). Most readers, editors, and reviewers prefer short, meaty, clear papers. Ensure that you have not repeated any information unnecessarily. Your writing will be more concise if you learn to recognize such repetitions. Finally, proofread the text for punctuation, spelling, and typographical errors.

### ➤ Incubate the manuscript in between revisions

You will not be able to do all this revising on one draft, so revise in stages. Do as much as you can on the first revision. When you no longer see anything to change, put the paper in a drawer again for a few days and let it "incubate." The longer you have let it incubate, the more likely you will see passages that you may want to change, recognize portions that need work, think of points to include, and notice those to omit or condense. Then you are ready to work on the second draft.

## 8.3 REVISING SUBSEQUENT DRAFTS

### ➤ Let some time elapse between revisions. Then check for content, logical organization, and style again

After you have waited a few days, you will be ready to look at your document with fresh, critical eyes. Start anew by rechecking your draft for content and logical organization, and then recheck for style, especially word location.

Be prepared for additional revisions, particularly if you catch yourself having to re-read certain passages. Stumbling across sentences is usually a good indication that more revision may be needed. In revising these questionable passages, check for poor sentence location, sentence structure, word location, and word choice as well as for noun clusters, unclear comparisons, lack of parallel form, change in key terms, lack of transitions, and use of nominalizations. You may also want to read your text aloud to help catch mistakes or awkward phrasing that may otherwise be missed during silent reading.

Revising large sections in a sitting will make your document smoother. You may not be able to do this in the first few revisions, but the more you revise, the more you will be able to read through the document in one go.

### ➤ Proofread your manuscript

Be aware that the process of revision can be endless. There will inevitably be something that you would like to change every time you read through, but you should not spend forever writing one report or paper. At some point, you have to stop revising. Keep in mind that the writing does not need to be perfect, just clear. When you no longer see anything to revise, proofread your manuscript.

In addition to revising and proofing your text, pay special attention to your Literature Cited/References section. Check to see that authors' names are spelled correctly, that authors' initials and citation page numbers are correct, and that references are accurate (see also Chapter 4).

### ➤ Ensure that you are submitting a complete and final version on time

When you have finished revising your document, make sure that you have included the final versions of the tables and figures. Recheck that you have followed the Instructions exactly. Pay attention to detail such as font (Times New Roman 10 or 12 point is the most preferred), margins (usually 1 in. all around), line spacing (usually double spaced), word count, page numbering, and line numbering if needed.

If there are no more corrections, make sure that the version you submit is really the final version and that it is complete. Ensure also that you turn in your report or article on time.

## 8.4 CHECKLIST FOR REVISING

**Individual sections**

Title:

- ☐ Is the title strong?
- ☐ Does it contain three to four important key terms?

Abstract:

- ☐ Have all necessary elements been included (question, experimental/study approach, results, interpretation/answer, significance)?
- ☐ Is the Abstract concise?

Introduction:

- ☐ Does the Introduction follow a funnel structure?
- ☐ Does the Introduction clearly state the overall question of the paper?
- ☐ Does the question follow the unknown?

- ☐ Has the question and experimental/study approach been stated in the last paragraph?
- ☐ Are all elements (known, unknown, question, experimental/study approach) clearly signaled?
- ☐ Did you ensure that the topic has not been reviewed?

Materials and Methods:

- ☐ Have all experiments and observations been described adequately?
- ☐ Are experiments and observations organized logically?

Results:

- ☐ Is the main finding presented in the first paragraph?
- ☐ Are there errors in factual information, logic, analysis, statistics, or mathematics?
- ☐ Are all figures and tables explained sufficiently?

Discussion:

- ☐ Is the overall interpretation of the results clearly stated in the first paragraph?
- ☐ Did you adequately summarize and discuss the topic?
- ☐ Has a clear conclusion been provided?
- ☐ Is the significance clearly stated in the last/concluding paragraph?
- ☐ Is the Discussion ordered in a way that is logical, clear, and easy to follow?

References:

- ☐ Have references been cited where needed?
- ☐ Are sources cited adequately, appropriately, and accurately?
- ☐ Are all the citations in the text listed in the References section?

**Style and composition**

- ☐ Are the transitions between sections and paragraphs logical?
- ☐ Are key words repeated exactly?
- ☐ Are the paragraphs and sentences cohesive?
- ☐ Has word location been considered?
- ☐ Did you check for grammar, punctuation, or spelling problems?
- ☐ Is the style concise?

## 8.5 EDITING SOMEONE ELSE'S MANUSCRIPT

Editing and evaluating the work of others is one of the best ways to reinforce familiarity with revising strategies. Such editing can also give the author a deeper understanding of how writing affects different readers and how manuscripts are edited professionally. In addition, such editing sets the stage for later peer review as professional scientists.

### ➤ Provide comments in writing

How you edit a peer's work depends on when in the writing process you are doing the editing. Early drafts should be evaluated primarily with respect to major components of the paper such as the research purpose, the main findings/answer, and the logical overall organization of the paper. Subsequent drafts should be edited for style and composition as well as for flow.

The least helpful comment to receive from a peer revising your work is "It looks OK to me." To be an effective editor, you need to be as specific as possible and point out both strengths and weaknesses. Point out particular places in the paper where revision will be helpful. Do not hesitate to note when something is unclear to you, scientifically or in terms of the writing. If you disagree with the comments of another person who has edited the manuscript, say so. Not all readers react the same way, and divergent points of view can help writers see options for revising.

Provide your comments in writing. Verbal comments are easily forgotten or confused. Write your comments either between the text lines or on the margins of the draft. Better yet, use the "Track Changes" option of Microsoft Word because this function will allow you to suggest wording and clearly show the author where and how to revise a document.

### ➤ Always treat the author with respect

Always treat the author with respect. Avoid snippy comments such as "So what?" Instead, make constructive suggestions and recommendations on how to improve and strengthen certain passages or on what else to add or omit from the document. If a passage reads well, point out this strength. If an argument is difficult to follow logically or does not make sense, raise objections politely or ask for explanations to clarify the argument.

## 8.6 CHECKLIST FOR EDITING SOMEONE ELSE'S MANUSCRIPT

### CONTENT

#### Purpose and interpretation

- ☐ Is the overall purpose of the paper and/or central question clear?
- ☐ Does the interpretation of the findings answer the overall question of the paper?

#### Support

- ☐ Is there sufficient evidence to support the answer?
- ☐ Is every paragraph and sentence in the paper relevant to the overall question?
- ☐ Are there portions of the text that could be omitted?

## Overall

- ☐ Does the paper advance the field?
- ☐ Does it provide interesting and important insights into the topic of interest?
- ☐ Have power positions been considered (especially in the Introduction, Results, and Discussion)?

## Individual sections

### Title:

- ☐ Is the title strong?

### Abstract:

- ☐ Does the Abstract adequately summarize the paper?
- ☐ Have all necessary elements been included (question, experimental/study approach, results, conclusion)?
- ☐ Is the abstract concise?

### Introduction:

- ☐ Does the Introduction clearly state the overall question of the paper?
- ☐ Does the question follow the unknown?
- ☐ Are all elements (known, unknown, question, experimental/study approach) clearly signaled?
- ☐ Did you ensure that the topic has not been reviewed exhaustively?

### Materials and Methods:

- ☐ Have all experiments and observations been described adequately?
- ☐ Are methods detailed enough that the study can be repeated by another trained scientist?

### Results:

- ☐ Has the main finding been clearly presented?
- ☐ Are there errors in factual information, logic, analysis, statistics, or mathematics?
- ☐ Are all figures and tables explained sufficiently?

### Discussion:

- ☐ Has the overall interpretation of the results been clearly stated?
- ☐ Did the writer adequately summarize and discuss the topic?
- ☐ Has a clear conclusion been provided?

## ORGANIZATION

### Overall organization

- ☐ Is the overall organization of the paper clear and effective?
- ☐ Are there unclear portions?
- ☐ Could the clarity be improved by changes in the order of the paper?
- ☐ Does the language seem appropriate for its intended audience?

### Individual sections

Abstract:

- ☐ Does the Abstract contain all key elements?

Introduction:

- ☐ Does the Introduction follow a funnel structure?

Materials and Methods:

- ☐ Are experiments and observations organized logically?

Results:

- ☐ Is the main finding presented in the first paragraph?
- ☐ Are all figures and tables labeled properly?

Discussion:

- ☐ Is the overall interpretation of the results stated in the first paragraph?
- ☐ Is the significance stated in the last/concluding paragraph?
- ☐ Is the Discussion ordered in a way that is logical, clear, and easy to follow?

References:

- ☐ Have references been cited where needed?
- ☐ Are sources cited adequately, appropriately, and accurately?
- ☐ Are all the citations in the text listed in the References section?

## STYLE AND COMPOSITION

- ☐ Are the transitions between sections and paragraphs logical?
- ☐ Are key words repeated exactly?
- ☐ Are paragraphs and sentences cohesive?
- ☐ Has word location been considered?
- ☐ Are there any grammar, punctuation, or spelling problems?
- ☐ Is the style concise?

- ☐ Are there any wordy passages?
- ☐ What other problems exist? ______________________________
______________________________

**OVERALL QUALITY**

- ☐ What are the paper's main strengths?
- ☐ What are the paper's main weaknesses?
- ☐ What specific recommendations can you make concerning the revision of this paper?

## 8.7 SUBMISSION AND THE REVIEW PROCESS

### ➤ Submit to only one journal at a time

Upper-level undergraduate students often are in the position of being able to publish their work, either in an undergraduate journal or in an academic journal as part of actual research they may have done in a laboratory. If you are submitting a paper for publication, follow the *Instructions to Authors* provided by your target journal to format your manuscript. Send your manuscript to only one journal at a time. Your paper will only be considered for publication if not submitted elsewhere concurrently. Electronic submissions are standard these days. Instructions for electronic submissions differ from journal to journal—follow them carefully.

### ➤ Ensure that you followed instructions explicitly

Recheck that you have followed instructions exactly. The biggest problem in submitting papers electronically is getting figures into the correct format and resolution. Here, it is especially crucial to follow guidelines and suggestions explicitly from the start so that figures and tables do not have to be remade at the last minute to fit the journal specifications. Usually, when you submit figures, you will receive electronic notification that the figure is in the correct format and has been accepted for submission.

### ➤ Wait for a response from the editor

Most journals will send an acknowledgment that the paper has been received. The *Instructions to Authors* or the acknowledgment you receive may say how long the journal will take to tell you the fate of the paper. The length of the review process depends very much on the journal. Expect a minimum of two weeks and up to six months to hear back with a decision on whether your article has been accepted or not. Do not submit the paper to any other journal until you get a letter of rejection from your target journal.

When the managing editor has received your manuscript, the editor will assign it to two to three anonymous, qualified reviewers. These reviewers will get back to the editor with specific comments and recommendations on the article. The editor then reads your paper and the comments of the reviewers. The editor may also comment on your paper, summarize the major comments of the reviewers, and state which reviewer comments should be taken most seriously in your revisions. Most important, the editor decides whether to accept, accept pending revision, reject with encouragement to resubmit, or outright reject your manuscript.

### ➤ If your manuscript is rejected, try, try again

If your paper gets rejected, read the reasons for rejection carefully. Various reasons for rejection can exist, including insufficient and unconvincing data, reviewer bias, incompatibility with the target journal, language issues, and lack of originality. Realize you are not alone. Up to 50% of articles submitted receive an initial rejection. More prestigious journals such as *Science* and *Nature* have even higher rejection rates (80%–90%). Relax. Then go to work on a revised version for a new journal that same day. The longer you wait, the harder it will be to get started again. Follow the suggestions of the reviewers when you revise your paper. Do not give up. Perseverance usually pays off.

### ➤ If your manuscript needs revisions, address every comment raised

If you have been invited to resubmit your paper after revisions, you need to address *all* criticism and concerns. If you have been asked to add additional data, you may have to go back to the lab bench to do so. Your resubmission should include a detailed list of responses to each suggestion and comment made by the editor and reviewers. Stay objective and professional; do not dismiss suggestions unless absolutely necessary. Editors and reviewers will not take kindly to that.

### ➤ When your manuscript has been accepted, celebrate

When you manuscript gets accepted, it is cause for celebration. So, get out the champagne. A party will serve as a great motivational basis for future teamwork on research and other manuscripts and documents.

When your manuscript is accepted, the text is set into type by the publisher. This process will take several weeks or months. In the meantime, you may be asked to assign copyright to the journal and to order reprints.

### ➤ Return proofs in time

It is likely that you will not hear from the journal for several weeks or months thereafter. Then the first proofs or galley proofs are sent out to the corresponding author. These proofs are a sample printing of your paper and will look very much the way the paper will appear in the journal. Examine the proofs or their electronic form closely and carefully. Look for errors, and make corrections and last-minute edits. At this point, you should not introduce any new material nor make any substantial changes. Also, do not make minor changes of wording or emphasis. Usually, only factual errors such as incorrect data in a table should be corrected now. If you need to make substantial changes, discuss this with the editor first. Return any future proofs or an electronic version thereof by the requested date as quickly as possible, usually within forty-eight hours. The next time you will see your paper it will be in print.

## SUMMARY

### FIRST REVISION

1. Check the first draft for content and content location.
2. Check logical organization of sections and subsections. Use checklists at the end of Chapter 7.
3. Revise for style only after you are satisfied with the content and organization.
4. Pay particular attention to key terms and transitions.
5. Condense where possible.
6. Proofread your manuscript.

### SUBSEQUENT REVISIONS

1. Let some time elapse between revisions.
2. Recheck the first draft for content and logical organization.
3. Recheck for style, and revise if needed.
4. Ask for comments and constructive criticism in writing.
5. Ensure that you are submitting a complete and final version on time.

### REVIEWING A MANUSCRIPT

1. Provide comments in writing.
2. Always treat the author with respect.

### SUBMISSION

1. Submit to only one journal at a time.
2. When your manuscript has been accepted, celebrate!

## PROBLEMS

### PROBLEM 8-1 Revision

**Revise the following Abstract. First, revise for content and content location, then check for logical flow and style, particularly signals, transitions, and key terms. Make constructive suggestions in writing.**

Using two different media (A and B), the bacterium *Enterobacter sakazakii* was found in 5 out of 50 powdered infant milk formulas and in 4 out of 25 milk powders. Although both media detected *Ent. sakazakii*, medium A proved to be about twice as sensitive in the organism's detection than medium B. *Salmonella* bacteria, the standard organism used for testing food products for the presence of *Ent. sakazakii*, were not detected in either medium.

Monitoring dry milk and food products solely for *Salmonella* is insufficient for the detection of *Ent. sakazakii*.

**PROBLEM 8-2 Revision**

**Write a section of a lab report, research paper, or review article. Then:**

1. **Revise your write-up.**
2. **Ask one of your peers to comment and edit the section.**

CHAPTER 9

# Sample Reports

THIS CHAPTER DISPLAYS:

- An example of a typical weekly lab report
- A sample project paper

## 9.1 GENERAL OVERVIEW

This chapter provides two well-written, full-length sample reports that follow key guidelines presented primarily in Chapters 7 and 8. The first exemplifies a typical lab report written during a weekly undergraduate laboratory course in biochemistry. The second has been composed as a final paper for a longer-term, inquiry-based project spanning about two-thirds of a semester-long microbiology lab course. Note that these are sample reports, and different schools and instructors may have somewhat different requirements on what to include or display.

Both reports follow the overall layout of typical scientific research articles, the IMRAD format (see Chapter 7), but the format of the weekly lab report is less stringent and also contains raw data and calculation steps. The final microbiology paper has almost all characteristics of a well-written scientific research paper in terms of format as well as style and tone of the paper. Annotations on the margins of each of these reports point out important passages and elements to be considered in writing your own reports.

## 9.2 SAMPLE WEEKLY LAB REPORT

**Example 9-1**

A clear title as well as the name of the author and the course number identify the lab report.

Tammy Wu

Bio 106L, 2008, week 6

LABORATORY 6: TIME COURSE OF β-GALACTOSIDASE INDUCTION

**Abstract**

This Abstract provides a general summary of the experiment, the approach used, and the overall finding.

In this experiment we measured the induction of the lac operon of *E. coli* when induced with IPTG (isopropyl thiogalactoside). When IPTG is added to the lac operon, the repressor–operator complex binding to the operon is dissociated and mRNA synthesis begins immediately. By adding the substrate ONPG (orthonitrophenyl galactoside), which is cleaved by β-galactosidase to generate the yellow compound *o*-nitrophenol, we could measure β-galactosidase activity by determining $A_{420}$ and $A_{550}$ absorbance. Thus, the time course of β-galactosidase synthesis was revealed.

**Introduction**

This introduction follows the funnel structure. It provides a general, brief background to provide context, and concisely states the purpose of the overall experiment as well as the general experimental approach in the last paragraph to let the reader know what to expect. (See also Section 7.7.)

β-galactosidase is an enzyme that catalyzes the hydrolysis of β-galactosides into monosaccharides. In *E. coli*, β-galactosidase is encoded by the *lacZ* gene, which is part of the *lac* operon. The operon is activated in the presence of lactose when glucose levels are low. Lac Z codes for the β-galactosidase monomer; active β-galactosidase is a tetrameric protein.

β-galactosidase is frequently used in genetics, molecular biology, and other life sciences as a reporter marker in various applications. In the β-galactosidase assay an active enzyme may be detected using the substrate X-gal, which forms an intense blue product after cleavage by β-galactosidase. β-galactosidase can also cleave ONPG (orthonitrophenyl galactoside), leading to the yellow product *o*-nitrophenol. Production of the enzyme can be induced by IPTG, which binds and releases the lac repressor from the lac operator. This release allows transcription to proceed.

In this experiment, we determined the time course of β-galactosidase synthesis by measuring ONPG cleavage during different time points after IPTG induction.

**Materials and Methods**

*IPTG Induction*

In this Materials and Methods section, experimental setups and execution are clearly described; the purpose for each step is also included at the beginning of each experimental subsection to provide context for the reader. (See also Section 7.8.)

To determine β-galactosidase enzyme activity for our experiment, we induced *E. coli* cells with IPTG. IPTG induction was stopped at various times by adding Z medium (containing β-mercaptoethanol; stops transcription) and toluene (allows ONPG to permeate cells). ONPG was then added and $A_{420}$ and $A_{550}$ measured to assay enzyme activity. Background absorbance due to cell density was obtained by determining $A_{600}$ of the grown *E. coli* culture twice. Background β-galactosidase activity (tube 2) was assessed by assaying *E. coli* cells to which ONPG had been added but without IPTG induction.

*Calculation of Raw Enzyme Units*

Subheadings are used in this Materials and Methods section, making it easy for the reader to find information.

To normalize for cell density at each time point, each $A_{600}$ value was determined by graphing $A_{600}$ before and after induction versus time, generating a standard curve (Figure 1, Table 1).

Raw enzyme units were calculated as follows:

Formulas and calculation steps are laid out clearly so the instructor can follow the student's analysis.

Units enzyme = 1,000 x $(A_{420} - 1.75 * A_{550}) / (t \times v \times A_{600})$
where $t = 54$ min for tubes 1–8,
$t = 14$ min for tubes 9–13, and $v = 0.5$ ml.

*Determination of Transcription and Translation Rates*

Background β-galactosidase enzyme units (tube 2) were then subtracted from each raw enzyme unit value, giving enzyme units from IPTG induction alone (Figure 2). By extrapolating this rate, transcription and translation rates were found (Table 2, Table 3a, Table 3b).

*Measurements and Data*

**Table 1 Cell Density Control Values**

| TIME | $A_{600}$ |
|---|---|
| –7 | 0.38 |
| 62 | 0.29 |

This lab report contains original raw data and calculations to let the instructor evaluate how and if the student arrived at the conclusions correctly. Note that for publication, grid lines should not be used.

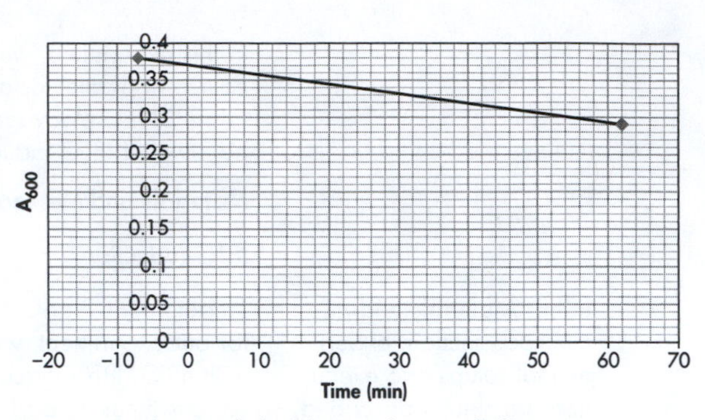


Figure 1. Background absorbance due to cell density.

**Table 2 Extrapolated $A_{600}$ Values**

| TUBE # | $A_{600}$ |
|---|---|
| 1 | — |
| 2 | 0.375 |
| 3 | 0.37 |
| 4 | 0.365 |
| 5 | 0.362 |
| 6 | 0.36 |
| 7 | 0.352 |
| 8 | 0.345 |
| 9 | 0.335 |
| 10 | 0.32 |
| 11 | 0.31 |
| 12 | 0.295 |

**Table 3a Calculated Enzyme Units**

| TUBE # | TIME AFTER INDUCTION (MIN) | ENZYME UNITS |
|---|---|---|
| 1 | | |
| 2 | –1 | 7.28 |
| 3 | 1 | 4.33 |
| 4 | 4 | 8.95 |
| 5 | 6 | 10.36 |
| 6 | 10 | 5.38 |

| | | |
|---|---|---|
| 7 | 15 | 17.81 |
| 8 | 20 | 10.31 |
| 9 | 30 | 28.89 |
| 10 | 40 | 27.79 |
| 11 | 50 | 27.3 |
| 12 | 60 | 36.08 |
| 13 | 70 | 27.0 |

**Table 3b Corrected Enzyme Units = Total Enzyme Units – Enzyme Units from Tube 2 (Background)**

| TUBE # | TIME AFTER INDUCTION (MIN) | CORRECTED ENZYME UNITS |
|---|---|---|
| 1 | | |
| 2 | | 0 |
| 3 | 1 | |
| 4 | 4 | 1.67 |
| 5 | 6 | 3.08 |
| 6 | 10 | |
| 7 | 15 | 10.53 |
| 8 | 20 | 3.03 |
| 9 | 30 | 21.61 |
| 10 | 40 | 20.51 |
| 11 | 50 | 20.02 |
| 12 | 60 | 28.8 |
| 13 | 70 | 20 |

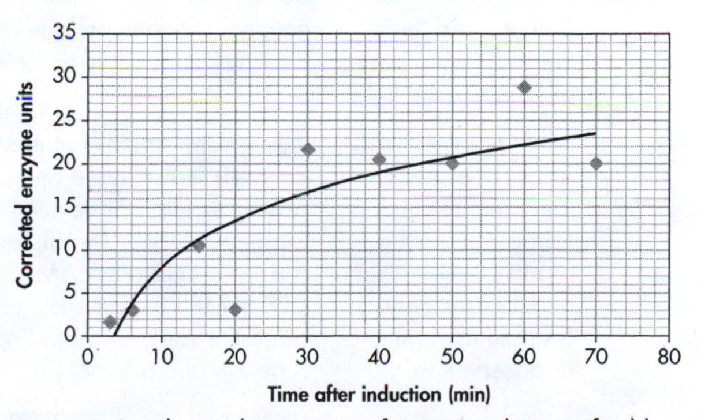


Figure 2. β-galactosidase activity after IPTG induction of wild-type *E. coli* cells.

**Results and Discussion**

A combined Results and Discussion section lays out the findings and speculates on their interpretation based on what was learned in class. Key results are described in the first paragraph. (See also Sections 7.9 and 7.10.)

In our experiment, according to the graph of β-galactosidase activity versus time of induction (Figure 2), enzyme activity was first noted at approximately 2 min after induction. Approximately 30 min after induction with IPTG, β-galactosidase activity started to level off and reached a saturation point. After 30 min, the assays all gave enzyme units over 20 but did not show the same increase rate as the previous times. This is likely due to the enzyme operating at maximum capacity, but could also be caused by the repressor binding to the operon and dissociation of cAMP, cessation of translation, or decreased viability of the enzyme (e.g., tetramer dissociation). All mechanisms could be mediated by the cell's own regulatory mechanisms for β-galactosidase activity.

This lab report lays out all limitations of the study.

Our experiment had some limitations. To graph a better curve, more samples at more time points would need to be taken, and experiments should be repeated. Aberrant values observed in our measurements and graph could have been due to nonhomogeneity of *E. coli* cells, or errors in adding reagents or in reading absorbance values.

The lab report explains the findings in context of what is known about the topic/has been learned in class, similar to a professional scientific research papers or term papers, which compare and contrast findings and their meaning with those of others in the field.

Because multiple copies of the lac mRNA exist, and would be translated simultaneously, tetramer assembly should be much faster than translation of the *lacZ* gene into the β-galactosidase monomer. The rate-limiting step in the start of enzyme activity, which took place about 2 min (120 s) after IPTG induction, was therefore assumed to be the protein synthesis of the enzyme monomer. Thus, the minimal rate of chain elongation of protein synthesis would be 1,170 amino acids (aa) in the monomer divided by 120 s, or 9.75 aa/s. Assuming that transcription and translation are coupled, the minimal rate of mRNA synthesis would be 1,170 x 3 / 120 = 29.25 bases/s.

Note that this lab report does not contain a concluding paragraph as the report is rather short and focuses only on a single experiment or observation.

If an adenyl cyclase mutant instead of wild-type *E. coli* cells would have been used, induction of the lac operon would not be seen because adenyl cyclase produces cAMP. The dissociation of the repressor–operon complex is cAMP-dependent; there is a CRP-cAMP binding site upstream of the RNAP and repressor sites, and binding of cAMP leads to transcription of the gene.

*(With permission from Tammy Wu)*

## 9.3 SAMPLE PROJECT PAPER

**Example 9-2**

| | |
|---|---|
| Author's name | Amanda Miller |
| Course | Microbiology |
| Date | 4/28/11 |

*An informative and complete title as well as the name of the author and course identify the lab report*

**Isolation and Characterization of *Bacillus amyloliquefaciens* from Facial Skin**

*The Abstract is concise and contains all essential elements and no citations. Elements are clearly signaled. (See also Section 7.6.)*

**Abstract**

The bacterial microflora is involved in a mutually beneficial relationship with the human body, where the microflora thrives in an ideal environment and the body gains protection against pathogens and receives aid in several processes. The purpose of this experiment was to identify one of the bacteria that inhabit the skin of the human body, specifically the facial skin, and to determine whether it is pathogenic, normal microflora, or a harmless transient bacterium. A sample of bacteria was taken from the cheeks, isolated, and analyzed with respect to morphology, Gram status, metabolic capabilities, antibiotic sensitivity, growth curve, and species identification with a Biolog assay. Biolog analysis revealed that the identity of the isolate was *Bacillus amyloliquefaciens*, a transient soil bacterium associated with plants. Characterization of the isolate highly consistent with the characterization of *B. amyloliquefaciens*, with the exception of some minor metabolic capabilities and morphology, indicating accurate species identification.

**Introduction**

*The first few paragraphs provide important background information for context. Information is paraphrased and sources are cited. (See also Sections 4.3, 4.6, and 7.7.)*

As soon as a human being is born, he or she comes into contact with microorganisms that colonize the body (Chiller et al., 2001). The human body and these microbes enter into a symbiotic relationship from which both gain some benefit. These microbes, which thrive and do not generally harm the host if colonized in the correct area or environment on the body, are known as normal human microflora (Willey et al., 2008).

The relationship between the microflora and the human body is beneficial to the microflora in that it provides an environment for them to grow and gain nutrients. This relationship is beneficial to the human host because the bacteria provide a source of protection against harmful biological agents, such as pathogenic bacteria, and aid in normal bodily processes. This relationship allows the correct balance of microflora to thrive on and inside the human body without causing it harm and instead increasing its health.

Sources are cited in parentheses at the end of the corresponding ideas, giving credit to the ideas and work of others. Placing references in parenthesis puts the emphasis on the information rather than on the authors of the corresponding works. The amount of information provided shows that the author has researched the topic sufficiently. (See also Section 4.3.)

The bacteria that inhabit the body are not uniformly distributed; some areas have more species of microflora than others, and not all microflora can exist on all parts of the body. The areas that they colonize on the human body are, usually, the skin and mucous membranes of the oropharyngeal tract, the gastrointestinal tract, and the female genital tract (Sullivan et al., 2001). The ideal environment for growth and survival differs among species, which is why each area of the body that supports normal growth has a different set of microflora (Tancrede, 1992). For example, the skin is generally dry and slightly cooler than the inside of the body, and has a slightly acidic pH (Chiller et al., 2001). The skin also has high concentrations of NaCl and can secrete inhibitory substances to prevent some bacterial growth (Willey et al., 2008). Just on the skin itself, varying conditions exist, especially in terms of moisture, pH, exposure, and capability to trap dirt (Evans et al., 1950). For example, the hands or fingernails may be more exposed to bacteria than skin on the face.

The microbes able to survive and benefit from using the skin as a host, therefore, differ from areas of the body which provide different environmental conditions. For example, the moist, often mechanically disrupted, neutral pH environment that is provided by the mouth will host different bacterial numbers and species than the skin, which is dry and acidic. Normal skin bacteria also differ from those in the GI tract. The microbes that may inhabit the large intestine are usually anaerobes, as they are not exposed to the external environment, while skin microbes are often found on superficial cells exposed to the air and therefore are frequently aerobic. Microbes on the skin and in the intestines are similar in that they must be continuously replaced, but for different mechanical reasons. Intestinal microbes are replaced as material moves through the intestine and transports them out. On the skin, microbes are replaced because the dead skin cells they adhere to are discarded frequently and are replaced themselves (Willey et al., 2008). Essentially, variances in the environmental conditions provided by the human body promote diversity in the microflora that uses it as a host.

Correct scientific nomenclature is used: italics for genus and species names; capitalizing genus but lowercasing species name. (See also Section 2.1.)

Several types of microbes are generally found at each site on the body. On the skin, Gram-positive bacteria predominate, including, but not limited to, *Staphylococcus aureus*, *Bacillus* species, *Streptococcus*, coagulase-negative bacteria like *S. epidermis* and *S. hominis*, *Micrococcus* species such as *M. luteus*, *Corynebacterium*, and *Propionibacterium* (Chiller et al., 2001). Although they are not the most common, the latter two are involved in causing acne lesions on the skin (Marples, 1974).

The unknown/problem/need is stated and signaled clearly. (See also Section 7.7.)

It is important to determine the distribution of microflora on the human body because of their role in protecting the body from pathogens and infection and in aiding internal processes such as digestion. Without the proper balance of microbes, the body will not be able to defend itself effectively and will not be able to break down materials needed for survival (Sullivan et al., 2001). This knowledge is especially important when considering the use of antibiotics. If the normal microflora is susceptible to the type of antibiotic being used to treat an infection, there is a chance that the normal microflora will be killed or significantly reduced (Edlund and Nord, 2000). Elimination of normal microflora diminishes its ability to protect the body by competing for nutrients, blocking the adherence of pathogens, producing toxins and inhibitory chemicals, and promoting the body's natural ability to produce antibodies and other defense mechanisms against pathogens (Chiller et al., 2001). The resulting lack of colonization resistance to opportunistic microbes such as *Clostridium difficile*, which can cause infection and disruption of the intestinal tract, can be extremely harmful (Sullivan et al., 2001). It is also important to study the distribution of microflora in order to understand the detrimental effects of a usually normal microbe growing in the wrong place. For example, if a bacterium that normally exists only on the skin gets into the bloodstream or a tissue that is usually sterile, that bacterium becomes a potentially harmful opportunistic pathogen (Willey et al., 2008).

In the last paragraph, the purpose of the experiment is clearly stated and signaled.

A good general overview of the experimental approach is provided. (See also Section 7.7.)

The purpose of this experiment was to isolate a single strain of bacteria from one site on the body—the skin on the outside of the cheeks, which is known to carry normal microflora—and to discover the identity of the isolate. To determine the strain of bacteria isolated at this site, several procedures and tests were performed, including phenotypic characterization (e.g., Gram status), analysis of metabolic capabilities, antibiotic sensitivity, growth curve, and ability to use varying carbon sources of chemical sensitivity assays in a Microlog plate.

This section is divided into subsections, each with its own specific heading, to help orient the reader.

Experimental descriptions provide sufficient details for other scientists to repeat the experiments; for example, every incubation step specifies temperature and time. (See also Section 7.8.)

**Materials and Methods**

***Isolation***

A sterile swab was dipped in sterile $dH_2O$ to moisten it before it was rubbed on the skin located on the outside of the cheek. Using sterile technique, a nutrient agar plate was inoculated by rubbing the sample swab across the surface of the plate, which was then incubated for 43 h at 37°C. Colony types were recorded, and one type was chosen for further isolation. A second nutrient agar plate was streaked for single colonies of the selected type and then incubated at 37°C for 24 h. A third nutrient agar plate was streaked again for single colonies of the selected type and was then incubated again at 37°C for 29 h. A fourth plate was re-streaked for single colonies and then incubated at 37°C for 48 h to ensure that the isolate was pure.

***Gram Stain/Morphology Identification***

Here, too, a step-by-step description of the experimental procedure is provided. The approach is specific enough to have others repeat the experiment, but does not contain any unnecessary details.

Sterile nutrient broth was inoculated with a single colony of the isolate AJM-3. After 24 h, a Gram stain of both the nutrient broth culture and the plate culture was performed. Using sterile technique, a smear from both cultures was placed within dime-sized circles on a glass slide (the plate sample was combined with a small drop of sterile water on the slide), allowed to dry, and was heat-fixed by passing the slide through the flame of a Bunsen burner. *S. marcescens* and *B. cereus* were also fixed on the slide as negative and positive controls, respectively. The slide was covered with enough drops of primary stain crystal violet to completely cover the samples and allowed to stain for 1 min. The stain was rinsed with $dH_2O$ before the slide was covered again with several drops of Gram's iodine mordant, which was also allowed to stand for 1 min.

The mordant was rinsed with $dH_2O$, and ethanol decolorizer was applied by running it across the slide for about 20 s, until most of the purple stain stopped running off of the slide. The slide was immediately rinsed with $dH_2O$ and counter-stained with safranin, which was rinsed with $dH_2O$ after 1 min. The slide was carefully blotted dry with bibulous paper and observed under the 100x oil objective for cellular morphology and Gram status, compared to the positive and negative controls. The presence of dark purple coloration was considered Gram positive, while pink coloration was distinguished as Gram negative. Colony morphology was identified by observation of the bacteria growing on an agar plate.

***Oxidase Test***

Control experiments (negative and positive) are included. Potential outcomes and their meanings are described, showing a good grasp of the material.

An oxidase test disk was placed on a clean glass slide using sterilized forceps. The disk was moistened by dipping it in sterile $dH_2O$ and placed back on the slide. A moderate amount of the isolate AJM-3 was sampled with a sterile inoculating loop and placed on the disk. This process was also performed on *E. coli* and *B. catarrhalis* as negative and positive controls, respectively. The disks were incubated at room temperature for 5 min and observed. Presence of purple color indicated cytochrome *c* oxidase was produced, while a lack of change in color was considered a negative result.

***Catalase Test***

An entire colony of AJM-3 was transferred onto a clean glass slide using an inoculating loop and sterile technique. One drop of 3% $H_2O_2$ was added on top of the sample and observed. The same procedure was performed on *E. coli* and *L. lactis* as controls. Presence of bubbles indicated that the bacterial species was positive for catalase production. Lack of bubbling indicated that the species does not produce catalase.

### ***Nitrate Reduction Test***

Every experiment described in the Materials and Methods section refers to a relevant experiment reported in the subsequent Results section, and vice versa, every result has a corresponding Materials and Methods section.

Tubes of nitrate broth were inoculated with AJM-3, *B. subtilis, E. coli,* and *L. lactis,* and one tube was left uninoculated as a control. The tubes were incubated at 37°C for 48 h. The cultures were removed from incubation, and 5 drops of Nitrate Reagent A were added, along with 5 drops of Nitrate Reagent B. The tubes were swirled and observed for color changes. A pinch of zinc dust was added with forceps to the tubes that did not change to red initially, and color was recorded again after swirling the tubes and letting them incubate for 10 min at room temperature. The cultures that initially changed color to red were determined to be positive for nitrate reduction. The cultures that changed to red after the addition of zinc were determined to be negative for nitrate reduction. Cultures whose color remained the same were classified as capable of complete denitrification.

### ***Mannitol Salt Agar Test***

AJM-3, *E. coli, S. epidermis,* and *S. aureus* were streaked, using sterile technique, onto an MSA plate divided into thirds. The plate was incubated at 37°C for 48 h and then observed for bacterial growth and change in media color. Presence of growth indicated that the species could tolerate the high salt concentration, and a change in the media color from red to yellow indicated that the bacteria could ferment mannitol. Lack of color change indicated that the bacteria were incapable of fermenting mannitol. AJM-3 was compared to the other three strains as controls.

### ***MacConkey Agar Test***

Here, too, potential outcomes and their meaning are described for every experimental approach described in the Materials and Methods section.

AJM-3, *E. coli, S. epidermidis,* and *S. marcesens* were streaked, using sterile technique, onto an MAC plate divided into thirds. The plate was incubated at 37°C for 48 h and observed for bacterial growth and color of bacterial growth. Presence of growth indicated that the bacteria were not inhibited by bile salts and crystal violet. Bacteria that appeared purple or red in color indicated that they were able to ferment lactose. Bacteria that appeared white or yellow indicated that they were unable to ferment lactose. AJM-3 was compared to the other three strains as controls.

### *Starch Hydrolysis Test*

An agar plate containing starch and iodine as an indicator reagent was inoculated with AJM-3, and *E. coli* and *B. subtilis* as controls, using sterile technique. The plate was incubated at 37°C for 48 h and observed for bacterial growth. Bacteria were determined to be positive for starch hydrolysis if they showed significant growth and formed a "halo" around the bacteria where the starch was hydrolyzed. Bacteria that showed little to no growth and had no "halo" surrounding the growth were considered negative for starch hydrolysis.

### *Antibiotic Sensitivity Test*

Although most Materials and Methods sections contain few if any references, citations can and should be added whenever appropriate, as at the end of this paragraph. (See also Section 7.8.)

A nutrient broth tube was inoculated with AJM-3 using sterile technique. Forty-five hours later, a sterile swab was dipped into the broth culture and rolled along the inside of the tube to rid the swab of excess moisture. A Mueller-Hinter agar plate was lightly swabbed in tight streaks and then rotated 1/3 of the way around. This process was repeated twice to create a lawn of bacteria, and the outside edge of the plate was swabbed again to ensure complete coverage. The plate was dried at room temperature. Antibiotic disks of erythromycin (15 μg), penicillin (10 units), rifampin (5 μg), streptomycin (10 μg), tetracycline (30 μg), and vancomycin (30 μg) were placed in a circular pattern on the plate using sterile forceps. The plate was incubated lid up for 48 hours. Zones of inhibition were measured and compared to the standards established by the Clinical and Laboratory Standards Institute to determine whether the bacteria were resistant, intermediate, or susceptible to each antibiotic (Madigan et al., 2009).

***Growth Curve and Generation Time***

To follow the cell growth for the isolate AJM-3, spectrophotometric determination of the optical density of a culture and the plating of a dilution series to count colony-forming units (CFUs) were used. In the spectrophotometric determination of the growth curve, a single colony of AJM-3 was inoculated into a sterile broth tube and incubated for 36 h at 37°C. Then, 1–5 ml of the overnight culture were added to 50 ml of sterile nutrient broth and incubated again for another 282 min. At this time, the culture was removed from incubation, and 2 ml were aseptically transferred to a clean cuvette. The culture was returned to incubation. A spectrophotometer set to 600 nm was blanked with a cuvette containing 2 ml of sterile nutrient broth, and the optical density of the 2 ml of culture was measured. This process was repeated every 20–30 min for 3–4 hours until optical density was measured at eight time points after the sterile nutrient broth was added. One optical density measurement was taken at a later time, 670 min after the sterile nutrient broth was added, for a total of nine time points.

In the CFU count determination of the growth of the cells, 1 ml of the broth culture of AJM-3 was removed 285 min after the sterile nutrient broth was added and transferred to a sterile microfuge tube. The sample was placed on ice and then diluted to $10^{-3}$, $10^{-4}$, $10^{-5}$, $10^{-6}$, and $10^{-7}$ dilutions of the original culture. These dilutions were then plated using the drip method; 10 µl of each dilution was pipetted in a line of drops across the top of the plate, allowed to run down the plate, and stopped before reaching the edge. The plate was allowed to dry, inverted, and incubated for 48 h. This procedure was repeated every 40–60 min for 3–4 h until four time points were taken following the sterile nutrient broth addition. One CFU time point was taken at a later time, 685 min after the sterile nutrient broth was added, for a total of five time points. After 48 h, the CFUs were counted in as many drips as possible, and the CFUs/ml were calculated at each time point.

To calculate the number of divisions or generations ($n$), the values for optical density at the first and last time points were used in the formula $n = (\log N - \log N_0)/\log 2$. This value and the total time ($t$) between the first and last time points were used in the formula for generation time, $g = t/n$.

***Microlog Species Identification***

A fresh nutrient agar plate was streaked with AJM-3 using sterile technique and incubated at 37°C for 21 h. A turbidometer was calibrated with an 85% turbidity standard. The turbidometer was blanked (adjusted to 100%) with the tube of IF-A. The IF-A tube was inoculated with AJM-3 by touching the tip of a swab to a single colony of the bacteria from the fresh plate and gently rubbing the swab on the bottom of the tube. The IF-A was recapped and inverted, and any extra particles were broken up with a new sterile swab. IF-A was placed in the turbidometer in the same position as when it was used to blank. Turbidity was measured at 98.5. The inoculum was poured into a sterile reservoir. A multichannel pipette was used to inoculate the MicroPlate with 100 μl of inoculums in each well. The plate was covered with its lid and incubated for 23 h.

The results of the assay were scored, comparing the A1 negative control to columns 1–9 and the A10 positive control to columns 10–12. Scoring was performed manually. For columns 1–9, "++" indicated very dark gray coloration, "+" indicated significantly darker coloration than the control well, "/" indicated coloration ambiguous in comparison to controls (very slightly darker or lighter), "–" indicated coloration the same as control, and "0" indicated a lighter coloration than the control. For columns 10–12, "+" indicates purple coloration similar to the positive control, "–" indicates purple coloration with less than half the color of the positive control, and "/" indicates ambiguity in the coloration in comparison to the positive control.

When entering the MicroLog plate data into the Biolog software, "++" and "+" were entered as positive, "0" and "–" were entered as negative, and "/" was entered as intermediate. The species ID, probability, and similarity identified through Biolog were recorded.

## Results

An overview paragraph, such as included here, is not always needed but helpful for long Results section such as this. (See also Section 7.9.)

The analyses of these discriminating tests were used to determine the identity of a microbe found on the human body. The results of morphology observation, Gram status, metabolic capabilities, antibiotic sensitivity, growth rate analysis, and a Biolog Assay were used to determine the specific species found on the site.

### *Isolation*

This Results section describes findings of the experiment in narrative form and points to relevant figures and tables, thus placing the emphasis on results rather than on data. (See also Section 7.9.)

The sample collected from the skin on the outside of the cheek, after incubation, yielded five different colony types. Type 1 was a large, cream-colored singular colony 6 mm in width, with irregular, raised, and undulate morphology. Type 2 included 30 small, pale yellow colonies evenly distributed over the plate, measuring about 0.5 mm in width, with circular, raised, and entire morphology. Type 3 colonies were also about 30 in number, were evenly distributed, and had circular, raised, and entire morphology. However, they were slightly larger than Type 2, measuring 1 mm in diameter, and were cream-colored. Type 4 included 40 colonies that were slightly less than 1 mm and were cream-colored, displaying circular, raised, and entire morphology. Type 5 included about 200 cream-colored colonies much less than 1 mm in diameter, with circular, raised, and entire morphology.

Findings are listed either chronologically or from most to least importance in this section, starting at the first paragraph.

Type 1 was selected for isolation. After streaking for a single colony of Type 1, the plate contained only the Type 1 colony type. Type 1 colony was restreaked a second and third time and was the only colony type on these plates as well, confirming a successful isolation of this specific bacterial species, temporarily named AJM-3.

### *Phenotypic Characterization*

The Results section is subdivided into subsections to ease readability; such subdivisions are often found in longer Results sections, such as the one here. The order of subsections is also kept similar between Materials and Methods and the Results sections. (See also Sections 7.8 and 7.9.)

Morphology, analyzed by Gram stain and streak plate observation, is helpful in finding correlations between the appearances of bacteria that are being compared. After further observation, the isolated bacteria showed irregular, raised, and undulate morphology consistent with previous observations. Gram staining revealed AJM-3 to be Gram-positive rods linked together in chains, sometimes two cells or four cells long.

***Metabolic Characterization***

Metabolic activities were measured in order to compare the specialized capabilities of the isolate. A summary of the metabolic activity of AJM-3 can be seen in Table 1.

Tables are numbered independently from figures. Table titles precede the table they refer to. (See also Section 6.5.)

**Table 1 Catabolic Activities of Unknown Bacteria AJM-3, Tested by the Oxidase Test, Catalase Test, Nitrate Reduction Test, Mannitol Salt Agar Test, MacConkey Agar Test, and Starch Hydrolysis Test.**

| TEST | AJM-3 |
| --- | --- |
| Cytochrome *c* oxidase production | Positive |
| Catalase production | Positive |
| Nitrate reduction | Reduction to NOs |
| Mannitol fermentation | Bacterial growth, no fermentation |
| Lactose fermentation | Fermentation, small amount of growth |
| Starch hydrolysis | Hydrolysis |

Information is ordered for the reader, i.e., independent variable on the left and dependent variables on the right. (See also Section 6.3.)

As seen in Table 1, AJM-3 produces both cytochrome *c* oxidase and catalase in its respiration processes. The culture of the broth turned red upon the addition of Nitrate Reagents A and B in the Nitrate Reduction Test, indicating that they reduced nitrate to nitrite, but are not capable of complete denitrification. The metabolic tests also indicated that AJM-3 is able to hydrolyze starch and that the species can ferment lactose but not mannitol. The mannitol test was somewhat ambiguous, as the media colors from the adjacent bacteria samples ran together in some places. The MacConkey Agar Test for lactose fermentation showed only a little bit of bacterial growth, but the color of the bacteria was purple, indicating that it was still able to ferment the lactose.

All findings have a corresponding Material and Methods section.

Results are not only listed but also briefly interpreted for the reader without going into any further discussion. (See also Section 7.9.)

***Antibiotic Sensitivity***

Figures and tables are mainly referred to in parentheses as they are supporting information. This way of presentation allows the author to place the emphasis on results rather than on figures and tables, which contain the data. (See also Sections 6.4 and 7.9.)

The Kirby-Bauer test, intended to address the specialized resistance abilities of an isolate, revealed that AJM-3 had different levels of sensitivity for each of the six antibiotics tested (Table 2)

**Table 2 Diameters of Zones of Inhibition and Assessment of Sensitivity of the Unknown Bacteria AJM-3 to Disks of the Antibiotics Erythromycin (15 μg), Penicillin (10 units), Rifampin (5 μg), Streptomycin (10 μg), Tetracycline (30 μg), and Vancomycin (30 μg) on Mueller–Hinton Agar plates**

| ANTIBIOTIC | AJM-3 |
|---|---|
| Erythromycin | 30 mm (susceptible) |
| Penicillin (Staphylococci) | 26 mm (resistant) |
| Penicillin (Others) | — |
| Rifampin | 19 mm (intermediate) |
| Streptomycin | 21 mm (susceptible) |
| Tetracycline | 17 mm (intermediate) |
| Vancomycin (Staphylococci) | 22 mm (susceptible) |
| Vancomycin (Enterococci) | — |

AJM-3 best resembled *Staphylococci,* based on the results of the Mannitol Agar Test in comparison to the control, and therefore was analyzed under the corresponding range for the zones of inhibition for penicillin and vancomycin.

AJM-3 was resistant only to penicillin but was susceptible to erythromycin, streptomycin, and vancomycin. AJM-3 had intermediate antibiotic resistance to rifampin and tetracycline (Table 2).

### *Growth Curve and Generation Time*

CFU counts and optical density measurements at specific time points were used to analyze the growth rate capabilities of the isolate. As seen in Figure 1, approximately 300 min after the overnight culture was added, the bacteria appeared to be in log phase of growth, which continued until optical density was measured at the last time point (about 700 min). The optical density growth curve data were estimated to have a better representation of the growth curve of AJM-3, as it had more data points and a more consistent pattern of growth over time, than the CFU growth curve data. The CFU growth curve data were inconclusive, and therefore generation time was calculated using the OD data.

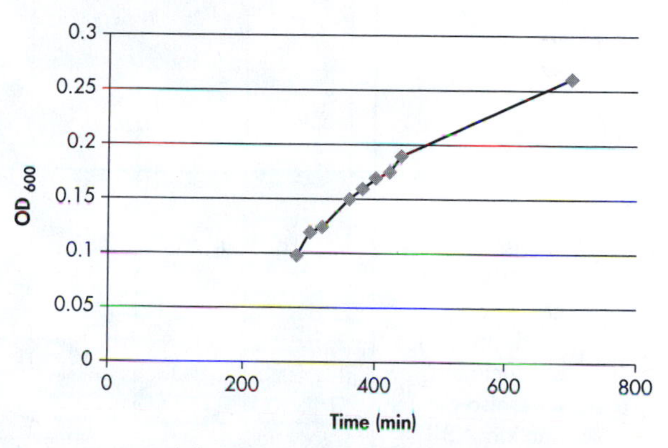


Figure 1. Optical density growth curve. Optical density was measured with a spectrophotometer at a wavelength of 600 nm at intervals after the addition of 5 ml of overnight culture of AJM-3 to 50 ml of sterile nutrient broth. Identified as log phase of the cell cycle.

In this graph, data are clearly visible and highlighted as they are drawn in the darkest entity. (See also Section 6.4.) The *x* and *y* axes have labeled variables and units. Tick marks are spaced out well, and lettering for them is large. The figure caption has been placed directly below the figure, and it contains a title as well as a legend/descriptive text.

Generation time for AJM-3, calculated from Figure 1, was 158.9 min.

### *Species Identification*

To comprehensively measure the sensitivity and utilization capabilities of the isolate, Biolog software was used to analyze a Microlog plate for AJM-3. The MicroLog plate used for bacterial identification was manually scored and entered into the Biolog system (Figure 2a and Figure 2b). A positive identification was made, indicating that AJM-3 was an isolate of the species *Bacillus amyloliquefaciens*. Probability of a correct identification was 0.957, and similarity was measured at 0.660.

In this presentation, chemical names are used; controls are included. (See also Section 7.8.)

| | 1 | 2 | 3 | 4 | 5 | 6 |
|---|---|---|---|---|---|---|
| A | Negative Control (faint gray) | Dextrin ++ | D-maltose ++ | D-Trehalose ++ | D-Cellobiose ++ | Gentiobiose ++ |
| B | D-Raffinose + | D-Lactose + | D-Melibiose / | Methyl-D-Glucoside ++ | D-Salicin ++ | N-Acetyl-D-Glucosamine ++ |
| C | D-Glucose ++ | D-Mannose ++ | D-Fructose ++ | D-Galactose / | 3-Methyl Glucose + | D-Fucose / |
| D | D-Sorbitol + | D-Mannitol + | D-Arabitol – | myo–Inositol + | Glycerol + | D-Glucose-6-$PO_4$ / |
| E | Gelatin ++ | Glycyl-L-Proline ++ | L-Alanine ++ | L-Arginine ++ | L-Aspartic Acid ++ | L-Glutamic Acid ++ |
| F | Pectin ++ | D-Galacturonic Acid ++ | L-Galactonic Acid Lactone 0 | D-Gluconic Acid ++ | D-Glucuronic Acid + | Glucuronamide + |
| G | p-Hydroxy Phenylacetic Acid – | Methyl Pyruvate + | D-Lactic Acid Methyl Ester / | L-Lactic Acid + | Citric Acid + | Keto-Glutaric Acid / |
| H | Tween 40 + | Amino-Butyric Acid / | Hydroxy-Butyric Acid 0 | Hydroxy-D,L-Butyric Acid – | Keto-Butyric Acid 0 | Acetoacetic Acid + |

The figure caption starts with a title followed by additional relevant information such as explanations for what abbreviations stand. (See also Section 6.4.)

Figure 2a. Manual scoring of Biolog assay from a GENIII MicroPlate, each well holding 100 µl of IF-A inoculating fluid containing AJM-3. "++", very dark gray coloration; "+", significantly darker coloration than the control well; "/", coloration ambiguous in comparison to controls (very slightly darker or lighter); "–", coloration the same as control; and "0", a lighter coloration than the control.

| | 7 | 8 | 9 | 10 | 11 | 12 |
|---|---|---|---|---|---|---|
| A | Sucrose ++ | D-Turanose + | Stachyose – | Positive Control (light gray) | pH 6 + | pH 5 + |
| B | D-Acetyl-D-Mannosamine ++ | N-Acetyl-D-Galactosamine – | N-Acetyl Neuraminic Acid 0 | 1% NaCl + | 4% NaCl + | 8% NaCl / |
| C | L-Fucose / | L-Rhamnose / | Inosine – | 1% Sodium Lactate + | Fusidic Acid – | D-Serine – |
| D | D-Fructose 6-$PO_4$ + | L-Aspartic Acid ++ | D-Serine – | Troleandomycin – | Rifamycin SV – | Minocycline – |
| E | L-Histidine ++ | L-Pyroglutamic Acid + | L-Serine ++ | Lincomycin – | Guanidine HCl + | Niaproof 4 – |
| F | Mucic Acid + | Quinic Acid ++ | D-Saccharic Acid – | Vancomycin – | Tetrazolium Violet – | Tetrazolium Blue – |
| G | D-Malic Acid – | L-Malic Acid ++ | Bromo-Succinic Acid ++ | Nalidixic Acid – | Lithium Chloride + | Potassium Tellurite + |
| H | Propionic Acid / | Acetic Acid ++ | Formic Acid ++ | Aztreonam – | Sodium Butyrate + | Sodium Bromate – |

Figure 2b. Manual scoring of Biolog assay from a GENIII MicroPlate, each well holding 100 µl of IF-A inoculating fluid containing AJM-3. Scoring for columns 7–9: "++" very dark gray coloration, "+", significantly darker coloration than the negative control well; "/", coloration ambiguous in comparison to the negative control (very slightly darker or lighter); "–", coloration the same as the negative control; and "0", a lighter coloration than the negative control. Scoring for columns 10–12: "+", gray coloration similar to the positive control; "–", gray coloration with less than half the color of the positive control; and "/", ambiguity in the coloration in comparison to the positive control.

## Discussion

The first paragraph of the Discussion contains the main findings and the answer to the research question. (See also Section 7.10.)

The species identified by the Biolog system, *Bacillus amyloliquefaciens*, is a bacterium normally found in the soil (Idriss et al., 2002). It is very closely related to *B. subtilis*, and was ultimately differentiated as a separate species not only because of DNA differences, but largely because of its ability to grow in certain environments (for example, in high concentrations of NaCl), its comparatively greater production of the hydrolyzing enzyme $\alpha$-amylase, which breaks down polysaccharides, and its ability to ferment lactose (Welker and Campbell, 1967).

*B. amyloliquefaciens* is normally associated with plants, and often serves as a biocontrol agent for the plants it colonizes. The bacteria colonize many types of plants and are especially important for the agricultural industry, as they protect the crops from disease and increase growth success (Arguelles-Arias, 2009). These bacteria produce antimicrobial dipeptides or cyclic lipopeptides (Yin et al., 2011). Such antimicrobial agents protect some plants against phytopathogens such as fungi, yeasts, and other bacteria (Yin et al., 2011). The capability of *B. amyloliquefaciens* to promote plant growth and protect from pathogens indicates its potential for involvement in developing biocontrol agents (Arguelles-Arias et al., 2009).

Although some *Bacillus* species are known to colonize human skin, *B. amyloliquefaciens* is normally only associated with plants, indicating that it was probably a transient bacteria on the cheek rather than a species of the normal microflora there. This is consistent with the findings from the initial swab of the cheek, which only yielded one colony of the bacteria that eventually was identified as *B. amyloliquefaciens*, as opposed to the four other colony types on the first plate, which were much more numerous.

Here and in the next few paragraphs, this Discussion compares and contrasts findings of this study with those of others in the field, again showing that the author has done his/her homework and researched

Previous literature indicates that the *B. amyloliquefaciens* are Gram-positive, rod-shaped bacteria that tend to form chains, as was indicated by the Gram stain and observations of cellular morphology of AJM-3 (Priest et al., 1987). The literature on colony morphology and coloring was fairly consistent with the present findings, with the exception that it described a circular form for the colonies rather than an irregular form (Yin et al., 2011).

the topic as the findings of the study are placed into context with other reports. Sources are referenced clearly in parentheses. (See also Section 4.3.)

This may have been due to the subjectivity of the observation rather than an actual morphological discrepancy.

Our data (Table 1) correlate well with results for testing methods described in the literature on *B. amyloliquefaciens*, which states that this species can hydrolyze starch, ferment lactose, and reduce nitrate (Welker and Campbell 1967). There is also correlation in the data stating that there is positive catalase and oxidase activity (Yin et al., 2011). However, the current study revealed that AJM-3 was negative for mannitol fermentation, which is not consistent with previous findings about *B. amyloliquefaciens* stating that the species does ferment mannitol as well (Priest et al., 1987).

Previous studies of antibiotic resistance in *B. amyloliquefaciens* often focused on different antibiotics than those in this study. However, it was emphasized that this species is highly resistant to penicillin, a result that, as seen in Table 2, was reflected in the Kirby–Bauer test (Dias et al., 1986). There is also evidence of limited resistance to streptomycin and tetracycline, which was also observed in AJM-3 (Dias et al., 1986).

Here, the Discussion provides potential explanations for observed differences between studies. (See also Section 7.10.)

The growth curve data in a study by Maubert et al. (2007). is somewhat consistent with the growth curve of AJM-3. It appeared to enter into log phase around 240 min after incubation started and continued until around 480 min, which is a shorter log phase than AJM-3 exhibited. However, the conditions of growth were different, and overnight culture was not added to the broth as it was to the culture of AJM-3, making it difficult to objectively compare the two sets of data (Maubert et al., 2007). The strain may also be different, which could potentially alter the growth curve. No generation time data were available for comparison for *B. amyloliquefaciens*, making the growth curve data least diagnostic in determining the correct species identification.

Inconsistencies and limitations of the study are also pointed out and discussed. Potential explanations are provided here as well. (See also Section 7.10.)

Finally, the Biolog data (Figure 2a) were compared to those of previous studies, revealing many consistencies, including positive carbon source utilization of cellobiose, maltose, fructose, mannose, lactose, maltose, glycerol, sorbitol, raffinose, salicin, trehalose, gelatin, and Tween 40 (Priest et al., 1987). Another notable consistency was the ability of AJM-3 (Figure 2b) and *B. amyloliquefaciens* to tolerate an environment of NaCl concentrations ranging from 1% to 8% (Priest et al., 1987).

One identifiable inconsistency was with the use of pectin as a carbon source. The Microlog plate indicated that it was used as a carbon source by AJM-3, identified as *B. amyloliquefaciens*, while the literature indicates that pectin may not be utilized by this species (Priest et al., 1987). This may be because different strains of the same species of bacteria may have slight variations in metabolic capabilities, or, potentially, just because of contamination of that well.

Another inconsistency found was between the MicroPlate data and the data from the Mannitol Salt Agar Test; the MicroPlate data showed that the bacteria can use mannitol as a carbon source in fermentation, which is contradictory to the negative finding in the earlier testing. This discrepancy is most likely attributed to misreading the MSA plate, which could have occurred if the media color associated with adjacent bacteria on the plate was mistaken for the media color surrounding AJM-3, given that the colors blended together somewhat on the plate.

Although there are a few small inconsistencies in morphology and metabolic capabilities, the high correlation between the testing results for AJM-3 and the established literature on *B. amyloliquefaciens* indicates that they are the same species. The inconsistencies are more than likely due to testing error or subjectivity rather than actual differences between the two species.

The last paragraph summarizes the main findings and provides the conclusion and significance of the study. (See also Section 7.10.)

Since the species identification appears to be correct, it is interesting to note that the bacteria were located on the skin of the face, an area very different from where it normally colonizes. This indicates that the bacteria are transient and that the skin may have somehow come in contact with a plant or the soil. Since the hands are a part of the skin constantly exposed to different environments and surfaces, it is possible that the soil and plant bacterium was first transmitted to the hands and then transmitted to the skin on the cheeks when the hand came in contact with the facial skin at some point. These findings also indicate how bacteria can be transmitted to an area of the body unintentionally, reinforcing the need for people to be aware of how harmless or harmful bacteria can colonize an area of the body if we are not mindful of what contaminated environments or surfaces we come in contact with unknowingly. *Bacillus amyloliquefaciens* is not a human pathogen, but its presence on the human skin shows how bacteria can be easily transmitted from one environment to another on the human body under the right conditions.

**Literature Cited**

References follow a required format consistently and precisely—in this case, references are listed alphabetically by last name. (See also Sections 4.3 and 4.4.)

Last names are followed by first name initial(s), then by year of publication, title of article, name of journal, volume, and page number.

Arguelles-Arias A, Ongena M, Halimi B, Lara Y, Brans A, Joris B, Fickers P. 2009. *Bacillus amyloliquefaciens* GA1 as a source of potent antibiotics and other secondary metabolites for biocontrol of plant pathogens. *Microbial Cell Factories* 8:63.

Chiller K, Selkin BA, Murakwaa GJ. 2001. Skin microflora and bacterial infections of the skin. *The Journal of Investigative Dermatology* 6(3):170–174.

Dias FFM, Shaikh KG, Bhatt YB, Modi DC, Subramanyam VR. 1986. Tunicamycin-resistant mutants *Bacillus amyloliquefaciens* are deficient in amylase, protease and penicillinase synthesis and have altered sensitivity to antibiotics and autolysis. *Journal of Applied Bacteriology* 60:271–275.

Edlund C, Nord CE. 2000. Effect on the human normal microflora of oral antibiotics for treatment of urinary tract infections. *Journal of Antimicrobial Chemotherapy* 46:41–48.

Evans CA, Smith WM, Johnson EA, Giblett ER. 1950. Bacterial flora of the normal human skin. *The Journal of Investigative Dermatology* 15:305–324.

Idriss EE, Makarewicz O, Farouk A, Rosner K, Greiner R, Bochow H, Richter T, Borriss R. 2002. Extracellular phytase activity of *Bacillus amyloliquefaciens* FZB45 contributes to its plant-growth-promoting effect. *Microbiology* 148:2097–2109.

Books are also included in the reference list and also follow a prescribed format. (See also Section 4.4.)

Madigan MT, Martinko JM, Dunlap PV, Clark DP. 2009. *Brock Biology of Microorganisms*, 12th ed. Reading: Benjamin Cummings.

Marples RR. 1974. The microflora of the face and acne lesions. *The Journal of Investigative Dermatology* 62(3):326–331.

Maubert ME, Hartz CB, Willson K. 2007. Identification of a *Bacillus amyloliquefaciens* (BA) strainable to bioremediate methyl tertiary butyl ether (MTBE) *in vitro* and *in situ*. *Journal of Young Investigators* 17(1).

Priest FG, Goodfellow M, Shute A, Berkeley RCW. 1987. *Bacillus amyloliquefaciens* sp. nov., nom. rev. *International Journal of Systematic Bacteriology* 37(1):69–71.

Sullivan A, Edlund, C, Nord CE. 2001. Effect of antimicrobial agents on the ecological balance of human microflora. Review. *The Lancet: Infectious Diseases* 1:101–114.

Tancrede C. 1992. Role of human microflora in health and disease. Review. *European Journal of Clinical Microbiology and Infectious Disease* 11(11):1012–1015.

Welker NE, Campbell LL. 1967. Unrelatedness of *Bacillus amyloliquefaciens* and *Bacillus subtilis*. *Journal of Bacteriology* 94(4):1124–1130.

Willey JM, Sherwood LM, Woolverton CJ. 2008. *Microbiology*. 8th ed. New York: McGraw-Hill.

Yin X, Xu L, Xu L, Fan S, Liu Z, Zhang X. 2011. Evaluation of the efficacy of endophytic *Bacillus amyloliquefaciens* against *Botryosphaeria dothidea* and other phytopathogenic microorganisms. *African Journal of Microbiology Research* 5(4):340–345.

*(With permission from Amanda Miller)*

CHAPTER 10

# Reading, Summarizing, and Critiquing a Scientific Research Article

THIS CHAPTER EXPLAINS:

- The information you will find in a typical scientific research article
- How to read a research article in a directed way
- How to compose a summary of a research article
- How to compose a critique of a research article

## 10.1 CONTENT OF A SCIENTIFIC RESEARCH ARTICLE

### ➤ Understand what information to look for and where to find it in a research article

If you are majoring in a scientific field, you will be reading articles published in academic and professional journals at some point. You might read these articles as part of a literature review you are required to submit, or your instructor may even ask you to write a summary or a critique of an article. If you are conducting research in a laboratory, you will also need to understand scientific research articles, which are also referred to as "research papers" (see also Section 7.1). Therefore, it is essential that you learn how to understand such articles, how to recognize their diverse elements, and how to summarize their content in your own words.

Research articles may seem overwhelming, especially to those who have little or no experience reading or writing this type of publication. Most likely you will not understand every detail. Reading and understanding "primary articles" or research papers (see Section 4.1 for a description of primary, secondary, and tertiary literature) is mainly a matter of experience and knowledge of the specific vocabulary of a field. However, a few simple tactics make this process much easier. Most important in this respect is to know what essential information you need to find and where to find this information in a research article.

Essential information you need to glean from an article includes the following elements:

- The overall purpose of the research
- The general experimental/study approach
- The key results
- The significance of the work

To know where to find this information, you need to understand how journal articles are structured and how to recognize their elements. Their structure and elements are discussed in detail in Chapter 7, and I encourage you to familiarize yourself with these components so that you can read research articles in a directed way. A brief overview is provided here:

### Abstract

- Provides a mini-summary of the paper and should contain all important information in form of known, unknown, research question/purpose, experimental approach, results, and significance. Thus, the Abstract will let you know what information the article covers.

### Introduction

- Gives relevant background information and introduces the problem/unknown in the field.
- Last paragraph—states the overall question or purpose of the paper and the general approach. (In addition, it may also state the main results and their significance.)

### Materials and Methods

- Tells the experimental/study approach. Most of the time this section is very technical and detailed. For an overview of the experimental/study approach, you can also refer to the end of the Introduction or to the Results section.

### Results

- Delineates all findings of the work.
- First paragraph(s)—states the main results of the research.
- Subsequent paragraphs—outline the general approach and findings for each step of the study.

### Discussion

- Discusses the key findings and their significance.
- First paragraph(s)—informs about the main results and their meaning; answers the overall research question of the paper.

- Last paragraph/conclusion—points out the importance and potential impact of the research; may also restate the main findings and point to the direction of future studies.

## 10.2 READING A RESEARCH PAPER

### ➤ Read research papers in a directed way:

- Gain an overview first.
- Clarify questions and unfamiliar terminology.
- Take notes.

Given knowledge of the general layout of a paper, you can read research articles in a directed way. This approach should provide you with a concrete idea of what this paper is about, what the main results are, and why they are important. You will be able to summarize or critique material only if you understand these elements, and you can compose a successful term paper, literature review, or even your own publication only if you know how to summarize background or overview material.

### ➤ First, read the Abstract, the Introduction, and the Conclusion

To make sure that you will not miss any of the key material, read the Abstract, Introduction, and Conclusion first. These sections should provide you with the overall purpose of the paper, the unknown/problem, the main findings, and the overall importance of the paper. See Chapter 7 to help you recognize important signals that identify these elements.

### ➤ Then, read through the entire paper

Look especially closely at the figures. Write down questions you have and terminology you do not understand. Look up simple words and phrases. This may help you answer many, if not most, of your questions. A medical or biological dictionary or the Internet will help you find important definitions. A textbook of the specific field may be an even better source because it provides more in-depth explanations.

### ➤ Re-read the paper again—this time for fuller understanding

To ensure that you caught and understood all important details of the article, re-read the paper again. If you need to write a summary or critique, take notes on each section using the questions in Section 10.3 or 10.4 as guidelines.

## 10.3 WRITING A SUMMARY OF A RESEARCH PAPER

### ➤ Think of a summary as an expanded version of the Abstract, written in your own words

Summarizing (or critiquing) a professional paper is a great way to learn more about primary scientific articles, writing research papers, and the research process itself. You will learn how researchers conduct experiments, make observations, interpret results, and discuss the impact of findings. Writing these summaries will also help you distinguish between more and less important information within a paper.

After reading the article carefully, specific questions you should be able to answer include:

- What is the overall purpose of the research?
- How does the research fit into the context of its field?
- What was the general experimental/study approach?
- What are the key findings?
- How are the reported findings different or better?
- What are the major conclusions drawn from the findings?
- What is the overall importance of the work?

To write a good and concise summary, think of your summary as an expanded version of the Abstract of the article. The following outline can serve as a guideline for writing a summary:

1. Begin your summary by describing the main question or purpose of the paper, and provide some brief context.
2. Explain how the authors approached the study.
3. Describe the key findings.
4. Briefly discuss the meaning of the findings.
5. Conclude by stating the overall impact of the research, and explain why you think the study is relevant.

You do not need to summarize each detail of the paper. Focus instead on providing the reader with an overall idea of the content of the article. Ensure that the terminology you are using is correct. Limit your summary to roughly one page.

**Example 10-1** ***Summary of* "Effects of Mountains and Ice Sheets on Global Ocean Circulation"***

The article "Effects of Mountains and Ice Sheets on Global Ocean Circulation" describes the impact of mountains and ice sheets on the large-scale circulation of the world's oceans. The researchers used a series of simulations with a new coupled ocean–atmosphere model (Oregon State University–University of Victoria model [OSUVic]), which combines ocean, sea ice, land surface, and ocean biogeochemical model components, to analyze the orographic effects of mountains and ice sheets on zonal wind stress, surface fresh water budgets, precipitation, atmospheric circulation, air temperature, and eddy kinetic energy.

The investigations show that the higher the mountain ranges, the less water vapor is transported from the Pacific into the Atlantic Ocean. As a result, the Atlantic is saltier and the Pacific is fresher because of the Rocky Mountains and Andes. In addition, deep-water formation is increased in the Atlantic but decreased in the Pacific, leading to more stratification in the Pacific.

Orography also affects zonal wind stress. Higher mountain ranges lead to increased zonal wind stress in the Southern Ocean, which is shifted southward by the presence of the Antarctic ice sheet. The combination of less atmospheric water vapor export from the Pacific to the Atlantic and southward-shifted Southern Hemisphere westerlies enhance deep-water formation in the North Atlantic and accelerate the Atlantic meridional overturning circulation.

The authors conclude that the configuration of mountains and ice sheets is key to determining the observed global circulation pattern of the ocean. Their findings may be important for future changes in this pattern due to climate changes—and consequently, important consequences for the carbon cycle—particularly if the Antarctic ice sheet is reduced or disappears.

**(Schmittner et al., 2011,* J. Climate, ***24****, 2814–2829)*

## 10.4 CRITIQUING A RESEARCH PAPER

### ➤ To critique a research paper, highlight the strengths and weaknesses of the research, its presentation, and interpretation

In a critique you are evaluating work done by another author. The purpose is to highlight strengths and weaknesses of the reported research approach, presentation, or interpretation in one to two pages. At the same time, you will deepen your knowledge of the topic. Understanding what to look for in critiquing a research paper will also aid you in composing and editing your own research paper.

In contrast to a summary, composing a research paper critique is subjective writing. It allows you to point out shortcomings, weaknesses, and limitations, and permits you to suggest how to improve the work, its presentation, or its interpretation. For example, in the summary you would answer the question "What major conclusions can be drawn from the findings?" whereas in a critique you would need to look at the conclusions critically, asking instead "Are the major conclusions drawn from the findings accurate and reasonable?"

Important questions that you should answer when you compose a critique for a research paper are listed in the following table, as are those for a summary:

| SECTION | SUMMARY | CRITIQUE |
|---|---|---|
| Introduction | • What is the overall purpose of the research?<br>• How does the research fit into the context of its field? | • Is the purpose of the study clearly stated?<br>• Are the ideas novel/original?<br>• Has relevant background information been provided? |
| Methods | • What was the general experimental approach? | • Did the authors use appropriate measurements and procedures? Is the authors' approach needed to answer the question of the paper? |
| Results | • What are the key findings? | • Did the authors obtain the expected results?<br>• Are results correctly interpreted and were all controls met?<br>• If some assumptions are made, are they realistic?<br>• Are figures and tables explained clearly?<br>• Are the key findings clear? |
| Discussion | • How are the reported findings different or better? | • Has the overall research question been answered?<br>• Is it clear how the research fits into the context of its field?<br>• Does the work make an important contribution to the field? |
| Conclusions | • What are the major conclusions drawn from the findings? | • Are the major conclusions drawn from the findings justified?<br>• What would you add to the conclusion?<br>• What would you say differently? |
| Importance | • What is the overall importance of the work? | • How important is the work? |

**Example 10-2**

**_Critique_ of "Effects of Mountains and Ice Sheets on Global Ocean Circulation"***

In the article "Effects of Mountains and Ice Sheets on Global Ocean Circulation," two features influencing the large-scale circulation of the world's oceans are investigated: mountains and ice sheets. Large-scale ocean circulation, which contributes to our climate system, is influenced by wind, buoyancy forcing, momentum fluxes, and orographic effects, such as mountain configurations and ice sheets. The latter two features have not been investigated in any detail prior to this study.

To assess the impact of orography, the researchers used a novel coupled ocean–atmosphere model (Oregon State University–University of Victoria model [OSUVic]), which combines ocean, sea ice, land surface, and ocean biogeochemical model components. Their work analyzes the orographic effects of mountains and ice sheets on zonal wind stress, surface fresh water budgets, precipitation, atmospheric circulation, air temperature, and eddy kinetic energy through four experimental calculations.

Their modeling shows that the higher the mountain ranges, the less water vapor is transported from the Pacific into the Atlantic Ocean. As a result, the Atlantic is saltier and the Pacific is fresher and deep-water formation is increased in the Atlantic but decreased in the Pacific. In addition, higher mountain ranges lead to increased zonal wind stress in the Southern Ocean, which is shifted southward by the presence of the Antarctic ice sheet. The authors conclude that the configuration of mountains and ice sheets is key to determining the observed global circulation pattern of the ocean.

The study does not take into account the influence of fresh water streams and currents on the salinity of the oceans nor gravity wave grads or any anthropogenic changes such as depletion of the ozone in certain areas. Furthermore, as they point out themselves, their horizontal resolution is relatively coarse, consisting of only 10 layers. Further simulations considering these additional factors and using a state-of-the-art climate model and a finer resolution would therefore be desirable. Without question, their findings may be important for future changes in the ocean circulation pattern due to climate changes, particularly if the Antarctic ice sheet is reduced or disappears.

*(Schmittner et al., 2011,* J. Climate, ***24**, 2814–2829)*

## 10.5 CHECKLISTS

### For a summary

- ☐ Did you understand what information to look for and where in the article to find it?
- ☐ Did you read the paper in a directed way?
- ☐ Did you find:
  - ☐ The overall purpose of the research?
  - ☐ The general experimental/study approach?
  - ☐ The key results?
  - ☐ The significance of the work?
- ☐ When composing your summary, did you:
  - ☐ Begin your summary by describing the main question or purpose of the paper and providing some brief context?
  - ☐ Explain how the authors approached the study?
  - ☐ Describe the key findings?
  - ☐ Discuss the meaning of the findings?
  - ☐ Conclude by stating the overall impact of the research and explain why you think the study is relevant?

### For a critique

- ☐ Did you understand what information to look for and where in the article to find it?
- ☐ Did you read the paper in a directed way?
- ☐ Did you find:
  - ☐ The overall purpose of the research?
  - ☐ The general experimental/study approach?
  - ☐ The key results? How are they different or better from other?
  - ☐ The significance of the work?
- ☐ When composing your critique, did you:
  - ☐ Provide a brief summary by describing the main purpose, approach, findings, and conclusion of the paper?
  - ☐ Highlight the strengths and weaknesses of the research, its presentation, and interpretation?
    - ☐ Is the purpose of the study clearly stated?
    - ☐ Are the ideas novel/original?
    - ☐ Has relevant background information been provided?
    - ☐ Did the authors use appropriate procedures?
    - ☐ Are results correctly interpreted and were all controls met?
    - ☐ Are figures and tables explained clearly?
    - ☐ Are the key findings clear?
    - ☐ Has the overall research question been answered?

☐ Does the work make an important contribution to the field?
☐ Are the major conclusions justified?
☐ How important is the work?

## SUMMARY

### WORKING WITH A RESEARCH PAPER

1. Understand what information to look for and where to find it in a research article.
2. Read research papers in a directed way:
   - Gain an overview first.
   - Clarify questions and unfamiliar terminology.
   - Take notes.
3. First, read the Abstract, the Introduction, and the Conclusion.
4. Then, read through the entire paper.
5. Re-read the paper again—this time for fuller understanding.
6. Think of a summary as an expanded version of the Abstract written in your own words.
7. To critique a research paper, highlight the strengths and weaknesses of the research, its presentation, and its interpretation.

## PROBLEMS

### PROBLEM 10-1 Summarizing a Paper

**Choose a recent, peer-reviewed research paper in your field of interest. Summarize the paper, and give a 5 to 10 minute verbal presentation.**

### PROBLEM 10-2 Critiquing a Paper

**Choose a recent, peer-reviewed research paper in your field of interest. Critique the paper, and give a 5 to 10 minute verbal presentation.**

CHAPTER 11

# Term Papers and Review Articles

THIS CHAPTER LAYS OUT:

- The difference between a term paper and a review article
- How to decide on a topic for a review
- How to write a term or review paper
- The different components of a term or review paper with annotated examples

## 11.1 PURPOSE OF REVIEWS

### ➤ Distinguish between review articles and term papers

Review articles, also known as review papers, provide an overview of the current state of knowledge, literature, and research on a particular topic. Thus, they are not original articles with new data but, instead, secondary sources representing a well-balanced overview of a timely subject. (See also Chapter 4, Sections 4.1 and 4.2, for discussion on primary, secondary, and tertiary sources.)

Term papers, which upper-level undergraduate and graduate students are often asked to write, can be viewed as a simplified version of a review article, the difference being that review papers deal more in depth with current topics published in primary literature, are usually longer than term papers, and are published as articles in scientific journals. Both types of papers do not just give an overall summary of the specific subject matter or theory to others. Instead, they both concentrate on comparing and contrasting information to present an argument.

Whereas a term paper requires students to locate information about a given topic to present their opinion and support thereof in a report, a review paper is a peer-reviewed scholarly article that presents or analyzes results

**TABLE 11.1 Comparison Between Term Papers and Review Papers**

| | TERM PAPER | REVIEW PAPER/ARTICLE |
|---|---|---|
| **Purpose** | Report on topic that also shows student's stand on topic<br>Serves as assignment for a grade | Peer-reviewed scholarly published article<br>Serves as secondary resource for others<br>Provides scientists with up-to-date information on a topic<br>Gives a critical evaluation of the original research findings of others |
| **Audience** | Instructor | Readers of academic journal |
| **Source material** | Mixed literature (primary, secondary, tertiary) | Mainly primary literature; some secondary sources |
| **Author** | Undergraduate, graduate student | Professional scientist, sometimes in conjunction with postdoctoral fellow(s) and student(s) |

of others published as recent, original research articles (primary sources) in academic journals.

Review papers therefore evaluate current knowledge in the field in order to synthesize information, present a new hypothesis, or create a novel model based on information and analysis from recent primary literature. These evaluations advance our understanding of the field and add a new dimension to a topic. Once published in a scientific journal, review papers provide scientists with the most up-to-date information as well as with the history and a critical evaluation of the topic. Review articles in turn then are also extremely useful as secondary literature sources when composing other scientific texts as they provide important overviews and background information on a topic across a field, including an analysis of any controversies that may exist among different studies. As the author of such an article, you can suggest which side of the conflict seems to be presenting the better arguments. You can also suggest possible next steps or propose a new model.

### ➤ Make the information understandable to scientists in related fields

It is particularly important that the information in a review article is understandable to scientists in other related fields. Thus, reviews should use simple words and avoid excessive jargon and technical detail. Figures and boxed material should summarize and generalize primary source data or highlight new ideas.

Although review papers deal with topics in more depth and at a more professional level than term papers, for the remainder of this chapter, term papers and review papers are treated as the same, given that their overall structure is largely identical.

## 11.2 DECIDING ON THE TOPIC

### ➤ Select a topic of appeal and importance

Some research topics are much easier to write about than others. When deciding on a topic for a review, it is best to select a subject of appeal and importance. Typically, issues that are of interest to the scientific community and for which sufficient source material exists are easier to cover than subjects that do not hold much interest or for which very little information can be gathered. Thus, issues of well-defined and well-studied areas of research generally give more fruitful topics.

To find out what is "hot" or "cutting edge," it is often helpful to read a couple of review articles from a variety of journals. Recent research articles might give you an idea of what is currently of interest in the field. Other ways to identify active areas of research are through reading editorials and letters to the editor.

There are two main approaches to choosing an area of research to write about in a review. One approach is to choose a point that you want to make and then select your primary studies based on this area of interest. Another approach is to familiarize yourself with a topic and arrange the material by theme or idea. That is, research the topic starting at general sources and work your way to specific sources.

### ➤ Look at tertiary sources first to get an idea, secondary to work on an outline, and primary to fill in the outline

Begin by looking at textbooks or Internet sources that are vetted through peer review or have refereed references backing up statements. These are generally considered to be tertiary sources and can provide a good overview of a topic. Next, narrow your search by reading up on topics that have been summarized in secondary sources such as in review articles. For both approaches, use secondary sources to come up with a general skeleton for your review paper. Then, use gathered, specific information from primary sources, which are first-hand accounts of investigations and include journal articles, theses, and reports, to fill in the skeleton of your outline.

For all your sources, keep good records from the beginning, particularly citation information. There is nothing more aggravating than having to rediscover where the source material originated from. See also Section 4.5 on how to manage references.

## 11.3 FORMAT

### ➤ For review papers and term papers, follow the overall structure:

**Title**
Abstract—not always required
**Introduction**
**Main Analysis section**
**Conclusion and/or recommendations**
Acknowledgments (only for articles for publications)
**References**

The standard organization of term papers or review articles differs from that of lab reports or research papers. The organization of term papers and review articles usually includes an introduction to the topic, a main section with headings and subheadings, a conclusion with recommendations for further research, and a lengthy reference section. Some term papers and review articles also contain an abstract. Depending on what type of article you write, your paper could follow a slightly different organization.

### ➤ Create an outline and subsections based on gathered information

To compose a logically structured review article, it is a good idea to create an outline. If you decide to compose your article without one, at least check the overall organization using a reverse outline during the revision stage. For a clear outline of your term or review paper, create subsections based on the information you have gathered from the literature. Give these subsections individual headings and subheadings, and then sort the information you have collected into the various subsections. Use bullet points or whole sentences under each heading or subheading. Subsequently, sort the information under each heading or subheading by similarities, contrasts, gaps in knowledge, and so forth.

### ➤ Write iteratively

As you are filling in your outline, re-read the source articles to ensure that you have not missed anything. Identify additional papers if needed, and re-sort your material again if necessary. Writing a review article is an iterative process. When you are satisfied with your outline, start writing the review article by linking all the ideas under each subheading. The following sections provide more details on how to logically organize the Introduction, Main Analysis section, and Conclusion of a review article.

In Example 11-1, an outline of a review is presented. Note that the author goes beyond simply listing facts. The author analyzes previously reported findings from various sources and presents the result of this analysis as a new hypothesis/model (listed in bold).

**Example 11-1** **Outline of review article**

*Title:* Species-specific cell recognition: Gamete adhesion of sea urchins

*Abstract*

Background: Importance of gamete adhesion in sea urchins—overview sentence

Problem: Lack of understanding of species-specificity

Topic: Need to fill the gap in understanding

Overview of report/article content (**novel hypothesis/ model of species-specific gamete interaction**)

*Introduction*

Background:

- Cell adhesion—overview
- Cell adhesion and interaction during fertilization—sea urchins
- Advantages of sea urchin model system

Problem: Lack of understanding of gamete interactions in different species

Topic: Need to fill the gap in understanding

Overview of content: **New model for gamete interaction**

*Main Analysis*

- Macromolecular interactions in sea urchin fertilization
- Species specificity in fertilization for diverse sea urchin species
- Bindin and sperm adhesion
    - Support for bindin's function
        - Evidence for molecular interactions
    - Bindin's receptor
- **Deducted novel hypothesis: model of interaction between bindin and its receptor**
    - Evidence through deletion mutants
    - Evidence through sequence analysis

*Conclusion*

- Brief summary of main findings
- **Summary of novel hypothesis and gamete interaction model**
- Projection to other species

*Acknowledgments*

*References*

In this outline, the authors progress from a general overview of cell adhesion to specific gamete interactions in sea urchin fertilization in the Introduction. For the Main Analysis section, the article is logically organized into discussions of macromolecular interactions in sea urchin fertilization, the species specificity of these interactions, and the molecular interactions of sperm adhesion. Importantly, the Main Analysis section ends with a proposed model for the interaction of the specific sperm and egg adhesive molecules based on the collective evidence from previous studies. This hypothesis fills the gap in knowledge of the exact molecular interactions during sea urchin fertilization.

The outline concludes by summarizing the main findings, restating the new, resulting hypothesis, and projecting the possibility of similar interactions during fertilization in other species.

## 11.4 TITLE

### ➤ Make the title short but informative

The title of a review paper should be short and clear. Short titles are preferred by most readers as they are more memorable. This preference is particularly pronounced for review articles, which are geared toward a wider audience. Ideally, keep your title to within thirty to fifty characters.

As for a research paper, the title of a review paper should also be informative to let the reader know immediately what the review is about and what main ideas will be discussed. It should contain the top three to four key ideas/key terms of the paper. One option to write a clear title for a review article is to start with the main overall topic followed by a colon and then one or more subtopics that best describe the contents of the review. Two examples of such titles are shown following.

**Example 11-2** **Title of review article**

**a)** Species-specific Cell Recognition: Gamete Adhesion of Sea Urchins

**b)** Proteins Crystallization: Successful Approaches

## 11.5 ABSTRACT

### ➤ Write the Abstract as a table of contents in paragraph form, and include:

- Background (optional)
- Problem statement (optional)
- Topic of review
- Overview of content

Not all review articles require an abstract. If they do, their abstracts differ from those of a research paper. Unlike abstracts for research papers, those for review papers are essentially tables of contents in paragraph form. They may contain some background information and/or a problem statement and should state the topic and purpose of the review. They usually include little, if any, methods or results. Although some abstracts end with a statement of significance, interpreting the main findings of a topic for the readers, most end with an overview sentence, listing what will be discussed in the document. Note, however, that such overview sentences are not appropriate in abstracts of lab reports or scientific research articles.

| **Example 11-3** | **Abstract of review article** |
|---|---|
| Overview of content | This paper describes how plants adapt to desert environments in differing locales. **We outline various ways that allow plants to collect and store water as well as a variety of characteristics that allow them to reduce water loss, and discuss particularly tolerance, evasion, and water storage/succulence in detail.** |

Longer abstracts for review articles contain the following elements: background, problem statement, statement of topic, and overview of content, as shown for our sea urchin fertilization example outlined in Example 11-1.

| **Example 11-4** | **Abstract of review article** |
|---|---|
| Background | Cell adhesion is essential for fertilization in most eukaryotes. Sea urchin gametes serve as an important model system of cell–cell adhesion in fertilization, particularly as only one adhesion, bindin, and its receptor appear to be involved. |
| Unknown/ problem | Although multiple reports have been published in the last few decades on specific interspecies gamete adhesions and fertilizations, a detailed understanding of the interspecies-specificity of sea urchin fertilization is lacking. |
| Topic statement Overview of content | This review analyzes the reported observations of interspecies cell adhesion and fertilizations and presents a hypothesis and novel model for the arising of species specificity among sea urchins. |

## 11.6 INTRODUCTION

### ➤ Organize the Introduction:

- Background
- Unknown or problem
- Topic of review
- Overview of content

The Introduction of a term or review paper should provide the big picture of the topic and grab the readers' attention. It should present some general background and state the central topic of the review. It should also make clear why the topic warrants a review.

After discussing the general background and aspects of existing research, the article should present any recent developments and describe what the problems with the existing research are and/or what is unknown. Subsequently, it should explain the overall topic and purpose of the review article. These statements may be followed by a description of the organizational pattern of the review. Do not make the Introduction longer than one-fifth of the review article.

### ➤ Phrase your topic statement carefully

The topic statement of a review paper is similar to the question or purpose of a research paper. Your topic statement will not necessarily argue for a position or an opinion; rather, it will argue for a particular perspective on the topic. The topic statement sets the tone for the rest of the paper and makes the importance of the research area clear. Sample topic statements for literature reviews include:

**Example 11-5** **Topic statement for review papers**

Various practices have been applied to counteract the spread of Africanized bees among European bees.

The development of new antibiotics has become the main focus of several biotech companies as more and more resistance develops.

Mathematical modeling of disease transmission is important to maximize the utility of limited resources.

In the following example of a complete introduction of a review article, the various components are indicated:

**Example 11-6** **Introduction of a review paper**

| | |
|---|---|
| Background | Global climate change is altering the geographic ranges, behaviors, and phenologies of terrestrial, freshwater, and marine species. A warming climate, therefore, appears destined to change the composition and function of marine communities in ways that are complex and not entirely predictable (1–5). |
| Unknown/ problem | Higher temperatures are expected to increase the introduction and establishment of exotic species, thereby changing trophic relationships and homogenizing biotas (6). Because organisms in polar regions are adapted to the coldest temperatures and most intense seasonality of resource supply on Earth (7), polar species and the communities they comprise are especially at risk from global warming and the concomitant invasion of species from lower latitudes (8–10). |

Background

Shallow-water, benthic communities in Antarctica (<100-m depth) are unique. Nowhere else do giant pycnogonids, nemerteans, and isopods occur in shallow marine environments, cohabiting with fish that have antifreeze glycoproteins in their blood. An emphasis on brooding and lecithotrophic reproductive strategies (11, 12) and a trend toward gigantism (13) are among the unusual features of the invertebrate fauna. Ecological and evolutionary responses to cold temperature underlie these peculiarities, making the Antarctic bottom fauna particularly vulnerable to climate change. The Antarctic benthos, living at the lower thermal limit to marine life, serves as a natural laboratory for understanding the impacts of climate change on marine systems in general.

Topic statemtent

Recent advances in the physiology, ecology, and evolutionary paleobiology of marine life in Antarctica make it possible to predict the nature of biological invasions facilitated by global warming and the likely responses of benthic communities to such invasions.

Overview

This review draws on paleontology, biogeography, oceanography, physiology, molecular ecology, and community ecology. We explore the climatically driven origin of the peculiar community structure of modern benthic communities in Antarctica and the macroecological consequences of present and future global warming.

*(With permission from Annual Reviews)*

## 11.7 MAIN ANALYSIS SECTION

**➤ Logically organize information within the Main Analysis subsections (similarities, contrasts, gaps in knowledge, etc.)**

It is difficult but critically important to organize the Main Analysis section logically. To do this well, you will need to summarize, generalize, and organize findings reported in primary sources while also comparing and contrasting them. You may present key original data or your extrapolated conclusions based on the results from the source articles. Avoid introducing new data unless absolutely necessary to make your point. Feel free to show data in figures and/or tables but do not provide experimental details. Ensure that you are not plagiarizing, however. Cite sources of information as needed. (See also Chapter 4, Section 4.6.)

**➤ Organize the Main Analysis section logically into subsections:**

- Chronologically
- Thematically
- Methodologically

Generally, this section of a term paper or review article can be organized chronologically, thematically, or methodologically. If you are organizing your article chronologically, you follow a logical timeline based on when relevant source articles were posted. Alternatively, you could examine the sources under the history of the topic. Such an organization would call for subsections according to eras within this history.

In contrast to the chronological presentation of topics, thematic reviews are organized around a topic or issue rather than around the progression of time. For example, as you deal with various levels of evidence pertaining to a question, your Main Analysis could move steadily downward in the level of inquiry from the organism, to the organ, to the cell, to the molecular mechanisms within the cell. Note that your discussion may still follow a chronological timeline, however.

If you are presenting your article using a methodological approach, you need to focus on the methodology presented in your source articles rather than on their scientific content. Accordingly, topics are organized by techniques or by methods or approaches.

You need to arrange the Main Analysis subsections of your review article based on its organizational structure. For a chronological review, subsections should relate to key time periods. For a thematic review, subsections should correspond to relevant themes or ideas, whereas for a methodological review, subsections should be based on description of different methods and approaches.

### ➤ Consider including other subsections as needed

Some reviews contain additional subsections outside those of your organizational structure, such as subsections on "Current Status" or "Future Directions." What subsections you need to include in the body may only become clear as the review evolves. Be creative if needed.

In the next example, a partial Main Analysis subsection of a review article is shown.

---

| **Example 11-7** | **Main Analysis section of a review paper** |
|---|---|
| Subheading | **Transport of pathogens** |
| Context<br><br>Problem/<br>unknown | There are economic and agricultural concerns raised by the possibility of intercontinental dust events enhancing the spread of plant and animal diseases. The limited genetic diversity of many modern crops increases the risk that a disease outbreak could quickly achieve worldwide distribution given that many of the plants are clones with identical susceptibility [35]. Of the microorganisms identified from African dust aerosols in three studies, 5–25% have the potential to be plant pathogens; that is, the genus or species identified is known to contain members that cause disease [13, 18, and 24]. Examples include *Bacillus pumilus*, which can cause bacterial blotch in peaches, and *Bacillus megaterium*, which can cause "wetwood" disease in trees. |

Transoceanic or intercontinental aerosol transmission of a livestock pathogen has not yet been reported. However, there have been reports of the foot-and-mouth disease virus (FMDV) being transmitted by aerosols from Germany to Scandinavia, and from France to England [36 and 37], which led to speculation that FMDV could be carried from Africa to Britain or South America via desert dust [38]. It has also been suggested that FMDV has traveled from China to Korea in Asian dust [33].

A hypothesis has been offered that infectious agents in African dust could be linked to widespread episodes of coral reef morbidity and mortality occurring across the Caribbean basin [7]. This hypothesis has been supported by the discovery of the fungus *Aspergillus sydowii*, causative agent of sea fan aspergillosis, in Sahelian soil [39] and African dust events sampled in the US Virgin Islands [26].

Although opportunistic human pathogens, such as *Aspergillus fumigatus*, *Aspergillus niger*, *Staphylococcus gallinarum* and *Gordonia terrae*, have been identified in African dust [12, 13, 18, and 24], there are no reports as yet of human infectious diseases related to LDD of desert dust. However, correlations between African dust events and increased incidences of asthma in the Caribbean have been proposed and confirmed [40 and 41].

*(With permission from Elsevier)*

## 11.8 CONCLUSION

### ➤ In the Conclusion section, summarize your topic, generalize any interpretations, and provide some significance

The Conclusion section is one of the main highlights of a term or review paper. It recaps your review and your main conclusions, recommendations, and/or speculations. You need to phrase this section with special care, summarizing and generalizing main lines of arguments and key findings. Discuss what conclusions you have drawn from reviewing the literature, and restate your interpretations. In addition, provide some general significance of the topic and results, and discuss the questions that remain in the area. Although this section is often longer than the Conclusion section of a research paper, try to keep it brief.

Example 11-8 provides an example of a Conclusion section for a review article, which ends with some recommendations based on the analysis of the review.

 **Example 11-8** **Conclusion of a review paper**

***Recommendations for management***

| | |
|---|---|
| Summary and Recommendation | We recommend the MARK method based on non-invasive genetic data, when possible, for future population size estimation of brown bears. This method appeared to be more reliable than the other field methods, had relatively small confidence intervals and costs one-third to one-fifth as much as the helicopter-based CMR method, in addition to the other advantages described above. Another approach is to combine various types of data from different sources and scales through powerful modeling approaches to obtain indirect estimates of population size (Wiegand et al., 2003). Such models can prove to be very useful in the conservation of rare and elusive animals like the brown bear where limited data is available. |
| Conclusions and recommendations | However, we would like to point out two main concerns about the genetic method. First, technical difficulties in the laboratory work, associated with low quality and low quantity DNA samples (Taberlet et al., 1999), should not be underestimated. Our own experience has shown that the DNA amplification success is unpredictable and depends on different factors, such as conservation conditions of feces in the field. For instance, the genotyping success rate seems higher for faecal samples collected in colder climates (P. Taberlet pers. comm.). Therefore, we recommend pilot studies to be conducted for every project aiming to estimate population size from non-invasive data (Hedmark et al., 2004 and Maudet et al., 2004). Secondly, the genetic method appeared to be sensitive to sampling intensity. It performed well in 2001, when 1.6 samples with usable DNA were collected per estimated bear, but less well in 2001, when the ratio was 0.7. We stress the importance of an adequate sampling. Future studies should aim at collecting 2.5–3 times the number of faecal samples as the "assumed" number of bears (considering that approximately 20–30% of the samples could not be genotyped). Depending on the estimates obtained after data analyses, the sampling effort could then be re-adjusted to obtain more reliable estimates. |

*(With permission from Elsevier)*

## 11.9 REFERENCES

**➤ Cite primary and secondary sources as needed**

Like in academic research papers, your interpretation of the available sources must be backed up with evidence. Therefore, cite primary and secondary sources where needed. The type of information you choose should relate directly to the review's focus. See Chapter 4 for more information on references.

## 11.10 CHECKLIST

### Overall

- ☐ Does the topic present something of interest to the field?

### Individual sections

Title:

- ☐ Is the title short (ideally, 30–50 characters)?
- ☐ Does the title contain the main three or four key terms?

Abstract:

- ☐ Is the topic stated precisely?
- ☐ Is an overview of the article provided?

Introduction:

- ☐ Does the Introduction have the following components?
    - ☐ Background
    - ☐ Problem or unknown
    - ☐ Topic or review
    - ☐ Overview of content

Main Analysis:

- ☐ Is your section logically organized?
- ☐ Did you analyze and interpret all information (rather than simply list facts and dates)?
- ☐ Does your paper present information objectively, including contradictory data and ambiguities?
- ☐ Are all figures and tables explained and labeled sufficiently?
- ☐ Do all the components logically follow each other?

Conclusion and/or Recommendations:

- ☐ Is the topic summarized and interpreted in the Conclusion section?
- ☐ Is the result of your analysis clear?

References:

- ☐ Have references been cited where needed?
- ☐ Are sources cited adequately and appropriately?
- ☐ Are all the citations in the text listed in the References section?

**Style and composition**

- ☐ Are the transitions between sections and paragraphs logical?
- ☐ Are paragraphs consistent?
- ☐ Are paragraphs and sentences cohesive?
- ☐ Has word location been considered?
- ☐ Have grammar, punctuation, and spelling been checked?
- ☐ Is the style concise?
- ☐ Are key terms consistent?
- ☐ Is the action in the verbs? Are nominalizations avoided?
- ☐ Did you vary sentence length and use one idea per sentence?
- ☐ Are comparisons written correctly?
- ☐ Are lists parallel?
- ☐ Have noun clusters been resolved?
- ☐ Are nontechnical words and phrases simple?
- ☐ Have unnecessary terms (redundancies, jargon) been reduced?

## SUMMARY

### WRITING TERM PAPERS OR REVIEW ARTICLES,

1. Distinguish between review articles and term papers.
2. Make the information understandable to scientists in related fields.
3. Select a topic of appeal and importance.
4. Look at tertiary sources first to get an idea, secondary to work on an outline, and primary to fill in the outline.
5. For review papers and term papers, follow the overall structure:
   - **Title**
   - Abstract—not always required
   - **Introduction**
   - **Main Analysis section**
   - **Conclusion and/or recommendations**
   - Acknowledgments (only for articles for publications)
   - **References**
6. Create an outline and subsections based on gathered information.
7. Write iteratively.
8. Make the title short and informative.
9. Write the Abstract as a table of contents in paragraph form, and include:
   - Background (optional)
   - Problem statement (optional)
   - **Topic of review**
   - **Overview of content**

10. Organize the Introduction:
    - Background
    - Unknown or problem
    - Topic of review
    - Overview of content
11. Phrase your topic sentence carefully.
12. Logically organize information within the Main Analysis subsections.
13. Organize the Main Analysis section logically into subsections either chronologically, thematically, or methodologically.
14. Consider including other subsections in the Main Analysis as needed.
15. In the Conclusion section, summarize the topic, generalize any interpretations, and provide some significance.
16. Cite primary and secondary sources as needed.

## PROBLEMS

### PROBLEM 11-1 Review Introduction

**Identify all the elements of the following review introduction. Is the introduction complete?**

Mitochondrial genomes differ greatly in size, structural organization and expression both within and between the kingdoms of eukaryotic organisms. The mitochondrial genomes of higher plants are much larger (200–2,400 kb) and more complex than those of animals (14–42 kb), fungi (18–176 kb), and plastids (120–200 kb) (Refs. 1–4). Although there has been less molecular analysis of the plant mitochondrial genome structure in comparison with the equivalent animal or fungal genomes, the use of a variety of approaches—such as pulsed-field gel electrophoresis (PFGE), moving pictures (movies) during electrophoresis, restriction digestion by rare-cutting enzymes, two-dimensional gel electrophoresis (2DE) and electron microscopy (EM)—has led to substantial recent progress. Here, the implication of these new studies on the understanding of *in vivo* organization and replication of plant mitochondrial genomes is assessed.

*(With permission from Elsevier)*

### PROBLEM 11-2 Abstract

**Is the following an abstract of a research article or that of a review article? Explain why.**

Interleukin 1 (IL-1), a cytokine produced by macrophages and various other cell types, plays a major role in the immune response and in inflammatory reactions. IL-1 has been shown to be cytotoxic for tumor cells (Ruggerio and Bagliono, 1987; Smith et al., 1990). To determine the effect of macrophage-derived factors on epithelial tumor cells, we cultured human colon carcinoma cells (T84) and an intestinal epithelial cell line (IEC 18) with purified human IL-1. Microscopic and photometric analysis indicated that IL-1 has a cytotoxic effect on colon cancer cells as well as cytotoxic and growth inhibitory effects on intestinal epithelial cells. Because IL-1 is known to be released during inflammatory reactions, this factor may not only kill tumor cells but also affect normal intestinal cells and may play a role in inflammatory intestinal diseases.

**PROBLEM 11-3 Abstract**

**Is the following an abstract of a research article or that of a review article? Explain why.**

In the last few decades Africanized honey bees have been spreading throughout South and Central America into the southern states of the US. During this spread, they have Africanized European bees largely through crossbreeding during mating flights, which almost always leads to the more aggressive Africanized bees. Various practices have been applied to counteract this trend. This paper analyses different practices used to counteract this trend and recommends the optimal approach to ensure pure European honey bees.

**PROBLEM 11-4 Review Title**

**Evaluate the following titles. Are they suitable as a good title for a review paper? Why or why not?**

a) Flying Fish: Locomotion in Water and Air
b) Spread of Africanized Honey Bees
c) Renaturation and Crystallization of the Recombinant Sea Urchin Protein Bindin

**PROBLEM 11-5 Conclusion**

**Is the following Conclusion section appropriate for a review paper? Explain why or why not.**

Sea urchin gametes exhibit a high degree of species specificity. However, some interspecific inseminations have been observed among certain sea urchin species. Species specificity has been shown to arise through interactions of external surface macromolecules on both eggs and sperm. The sperm cell's primary adhesive substance, bindin, is found in the acrosome, and the cognate ligand on the egg is found in the egg's vitelline layer. Interactions between these macromolecules determine successful gamete recognition, adhesion, and fusion. While the egg ligand has not been studied in detail, analysis of the species specificity of bindin deletion mutants and a comparison of the protein sequences of these mutants among several sea urchin species suggests that the macromolecules function like a lock and key system whereby certain repeated sequences on the bindin molecule determine species specificity.

**PROBLEM 11-6 Conclusion**

**Is the following Conclusion section appropriate for a review paper? Explain why or why not.**

Many ectotherm animals hibernate in winter. Here we reviewed the hibernation of land snails. Their hibernation is triggered by decreasing day length (from 16 to 12 hr), low temperatures, and a decline of the water.

During hibernation, snails form an epiphragm to seal off their shell, stop feeding, reduce their oxygen consumption by half, and their heart rate by 90% or more, pending the temperature. We compared the evolution of these characteristics and the underlying differences to the hibernation of insects and vertebrates.

**PROBLEM 11-7 Main Analysis Section**

**For the Conclusion section of a) Problem 11-5 or b) Problem 11-6, construct a potential outline of the Main Analysis section you would expect to see in the review.**

CHAPTER 12

# Note Taking and Essay Exams

IN THIS CHAPTER YOU LEARN:

- How to take and study notes effectively
- How to prepare for an essay exam
- How to interpret essay questions
- How to answer essay questions
- Alternative strategies for answering essay questions
- Time management tips for answering essay questions

## 12.1 GENERAL OVERVIEW

Taking an exam—and particularly an essay exam—may inspire fear, especially when you may be under time pressure to complete the task. However, exams also provide an opportunity for you to show your understanding of the topic clearly. Many instructors like short answers and essays in addition to multiple choice questions because writing out an answer challenges students to create a response rather than to simply select a response, showing their understanding of a subject matter as well as their abilities to reason. In addition, written answers allow an instructor to assess writing abilities. One of the key requirements to do well on such exams is to take and study notes effectively.

## 12.2 EFFECTIVE NOTE TAKING

Learning how to take notes effectively is important and helpful for several reasons:

- To have a record of information to study and review
- To help you remember information
- To help you focus on the topic at hand
- As proof of what was discussed and said

This skill is not just useful to have as a student but also later in the professional world.

### ➤ Write notes down on paper

Although some people are good at taking notes on a tablet or on their computer, physically going through the motion of writing down the text by hand will allow most people to remember the notes better. It will also allow you to doodle, draw figures, make side notes, and so on to enhance your learning. This form of note taking is also less intrusive for the instructor and the preferred and more respectful way of note taking in most of the professional world. The next example shows a page of notes that have been well written.

**Example 12-1**

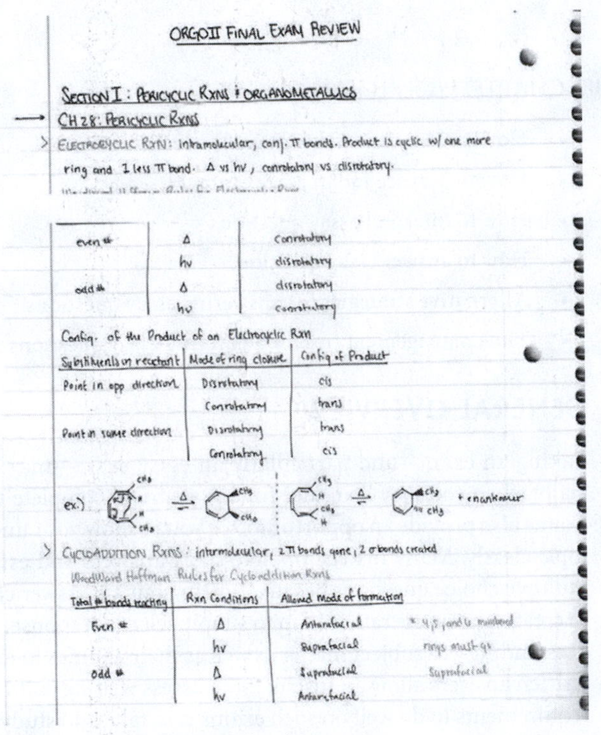


Figure. Example of notes taken by a student.

*(With permission from Nora Chov)*

### ➤ Use shorthand, abbreviations, and symbols

These skills are particularly useful when you are listening to a fast speaker. Make your notes as brief as possible; do not write full sentences. Prefer jotting down a word over a phrase over a sentence. Use bullet points. If you do not have the time to learn shorthand (a method of handwriting quickly using simple strokes, abbreviations and symbols), develop your own list of abbreviations and symbols. For example, place a star or exclamation mark next to something important, or draw out relationships using arrows. A list of common abbreviations and symbols follows.

## COMMON ABBREVIATIONS FOR NOTETAKING

| | |
|---|---|
| ~ | about |
| α | alpha |
| β | beta |
| & | and |
| + | and |
| … | and so on |
| @ | at |
| b/c | because |
| Δ | change |
| Ch | chapter |
| o | degree |
| diff | different |
| $ | dollar or money |
| ex | example |
| ' | feet |
| " | inches |
| 4 | for |
| ↑ | increase |
| ↓ | decrease |
| < | greater than |
| > | less than |
| ! | important |
| * | important |
| *** | very important |
| → | leads to |
| max | maximum |
| min | minimum |
| # | number |
| pg | page |
| % | percent |
| ? | question |
| rxn | reaction |
| ∝ | related to |
| = | the same as, identical |
| ≠ | different from, does not mean |
| Σ | sum |
| w/ | with |
| w/i | within |
| w/o | without |
| wrt | with respect to |
| ∴ | therefore |
| x | times |
| 2 | two, too, to |

### ➤ Organize and structure your notes as you go

Leave a large left margin for questions and key words—these can be filled in during and/or after the lecture—and some space at the bottom of the page for a summary (Cornell method). Also leave plenty of white space in the text so you can fill in additional information later on if needed. Structure notes in a hierarchy (similar to an outline). Underline the key points to help you find them and remember them easier later on. Support key points and subtopics with bulleted lists as appropriate. Indicate relationships between ideas using arrows and other visuals.

### ➤ Draw pictures

Use visuals to speed up your note taking and to increase your memorization. Visuals aid the brain in processing information. Visuals can be drawings, graphs, or diagrams. Use color, different font sizes, etc. to indicate relationships. Some of the most effective note-taking methods are mind maps, which allow you to develop connections easily in a visual environment (see also http://www.mindmapping.com and http://www.inspiration.com/visual-learning/mind-mapping). Figure 12.1 shows an example of a mind map.

### ➤ Pay attention to content

You should learn to identify the key points of what is being said. You almost never need to take down every word your instructor says. Rather, pay attention to the cues indicating what is important. Write down the information given by these cues. Such cues include:

- The instructor stating that "This will be on the exam" or "You should know . . ."
- Anytime the instructor repeats information.
- Everything on a Microsoft PowerPoint slide.
- Definitions that are given—write them down word-for-word.
- Anything the instructor says very slowly.
- Anytime the instructor talks with more emphasis.

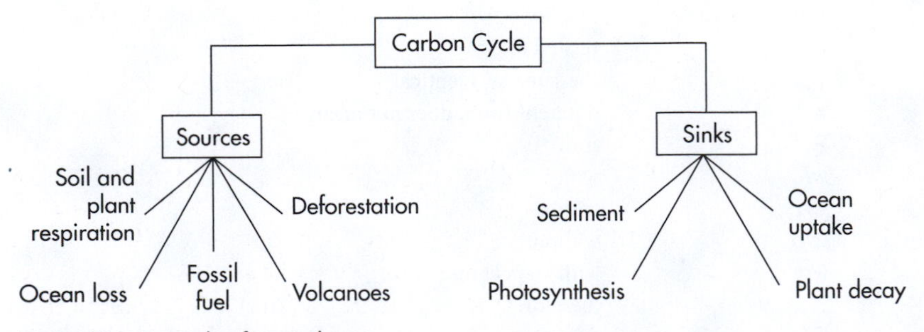


Figure 12.1 Example of a mind map.

- Information following words such as *first, second, third; especially/ particularly, importantly; however/but; because, therefore.* These indicate relationships between ideas and are often used as exam questions.
- Any examples the professor offers.

### ➤ Go over your notes right after the lecture

This may be the most important part of note taking. Take the time to *translate* your notes if needed. Ensure that you can read and remember any abbreviations and symbols; otherwise, spell them out as long as it is fresh in your memory. Indicate questions on the left-hand margin. If needed, fill in gaps and details using the textbook or check with classmates or the teacher. If you recorded the lecture, you can go back over the recording and find information you might have missed. Consider typing up your notes.

### ➤ Study your notes

Although re-reading notes, highlighting, underlining, and summarizing are common practices to help people in studying information, more effective study techniques include taking breaks and spreading out your studying as well as taking practice quizzes and tests. For examples of popular note taking formats, see https://asc.calpoly.edu/ssl/notetakingsystems. It may be worthwhile to determine which method works best for you.

## 12.3 PREPARING FOR AN ESSAY EXAM

### ➤ Prepare for an essay exam

The best and most important preparation for an essay exam is to study and understand the material. Commit key events, facts, and names to memory. Go to lectures, do the reading that is required, and learn the material as it is presented. Do not try to "cram" just before the exam. You will not be able to remember and know all the material at the depth required for an essay test by just trying to memorize as much as possible the night before.

You can prepare for an essay exam by:

- Taking good lecture notes and reading complementary material.
- Organizing and prioritizing information into topics.
- Practicing writing answers based on a study guide if provided.
- Studying notes and review questions.
- Discussing the study material with other students.
- Reviewing homework questions.
- Anticipating essay question(s) and practicing writing their answers. For example, you could turn the headings and/or subheadings in your textbook into questions and ask yourself to describe, explain, compare, or contrast these topics. Titles and subtitles introduced or favored by your instructor may be of particular importance for

anticipating potential essay questions. You could also have other students prepare essay questions for you.

- Knowing the grade breakdown for an exam as well as the point distributions within it. This information will tell you how much importance and attention to give to specific essay questions and how much time to budget for answering each question on an exam.

## 12.4 ANSWERING AN ESSAY QUESTION

To answer essay questions successfully, you need to:

1. Read and understand the question.
2. Plan your response: Jot down ideas/brainstorm and make an outline.
3. Write the first sentence as an overview sentence.
4. Logically organize and present your ideas.
5. Proofread your answer.

### ➤ Read and understand the question

Read each question carefully. If the question is not clear to you, ask your instructor for clarification.

Questions consist of three important elements:

- **Topic**, which identifies the subject.
- **Limiting words**, which indicate the subtopic or focus of the question and answer.
- **Task words**, which instruct you what to do.

To present a relevant answer, you need to analyze the question based on these three elements. This analysis will also allow you to structure the Introduction for the answer.

**Example 12-2** **Analysis of essay question**

*Essay question: Select one aspect of the effects of solar flares on Earth and discuss it in depth in relation to one level of reasoning.*

**Topic:** effect of solar flares on Earth

**Limiting words A:** one aspect only of the subject

**Task words 1:** Select . . .

**Limiting words B:** (your chosen aspect) in relation to one level of reasoning.

**Task words 2:** discuss in depth . . .

Note: you are being asked to choose *one aspect only* of the subject and discuss it in relation to *one level only* of reasoning.

In your essay answer, do not include all you know on the topic or subject. Instead, focus on the limiting words. If you have to incorporate specific information in your essay or restrict yourself to a particular set of conditions (e.g., "Restrict your answer to altitudes of more than 1,000 m"), make sure you cover the task and consider the limitations in your answer. If your instructor has set a word limit for the answer of the essay question, you should not go over that limit to avoid losing points.

Pay particular attention to the task words used in the question. You may find it helpful to circle or underline them. Learn to distinguish between the various task words and what specifically they are asking you to do (see Table 12.1).

**TABLE 12.1 Task Words**

| | |
|---|---|
| Argue | Give an opinion and defend it with evidence. |
| Define | Give the meaning of an idea and provide an appropriate example. |
| Describe | Present the main points and aspects of a subject. |
| Discuss/explain | Present the main points, facts, and details of a subject or topic and provide reasons. |
| Compare and contrast | Discuss similarities and differences. |
| Contrast/differentiate/distinguish | Present only the differences. |
| Criticize/evaluate/assess/analyze | Provide an analysis; discuss the strengths and weaknesses, pros and cons, advantages and disadvantages, supported by facts. |
| Interpret | Analyze and explain a topic using facts and reasoning. |
| Illustrate/diagram/draw | Provide a drawing or use a figure or diagram to explain something. |
| Justify/prove | Provide evidence and reasons for a statement. |
| List/identify/outline | Present a list of the main ideas and explain. |
| Review | Present a survey of a topic and evaluate evidence. |
| Summarize | State the main points and interpretations briefly. |
| Trace | Outline or discuss based on a pattern such as chronologically. |

### ➤ Plan your response: Jot down ideas/brainstorm and make an outline

After you have read and analyzed the question, note the expected length of the essay. For short essay answers, you will have less time and space, and a paragraph will almost always suffice. Longer answers will consist of multiple paragraphs and require more planning and thought. Note: Because you will usually be given more time for these types of questions, your instructor will probably expect higher quality in your writing. Therefore, drafting a *brief outline* of the answer will be particularly important for long essay questions. Making a *quick diagram* can also help you focus your thoughts. Use a pencil if possible to make changes and corrections more easily.

Long-answer essays may ask you to compare and contrast two ideas, or they may ask about several basic elements of a topic. Answers to such questions should be three to five paragraphs long and follow the following general format:

**Introduction** (first sentence or paragraph)—to introduce the topic/main ideas
**Body** (1–3 paragraphs)—to fill in the details
**Conclusion** (last paragraph)—to summarize, provide a thesis or significance

Start by jotting down ideas, then put the information in logical order—for example, chronological, most to least important, or pro and con. At this point, do not worry too much about proper sentence and paragraph construction, but be sure to write down all pertinent information in an organized fashion. The logical organization of your ideas will be essential for earning a good grade.

Continuing on our prior example, your notes, for instance, may read as follows:

---

**Example 12-3** **Planning your essay**

*Essay question: Select one aspect of the effects of solar flares on Earth and discuss it in depth in relation to one level of reasoning.*

**Notes:** Solar flares affect climate conditions on Earth.

Examples of effects:

- Increase in electromagnetic waves
- Increase in temperature

Increase in temperature:

- Mechanism
- Length
- Acc. to X theory, increases in regional temperature changes occur but will not affect overall climate change

---

### ➤ Write the first sentence as an overview sentence

When you are ready to compose your answer, state your main point in the first sentence, thus introducing the topic. This is best done by writing an overview sentence that repeats the key words found in the question and provides an overview of the details to come. For our prior example, your opening sentence may read as follows:

---

**Example 12-4** **First sentence**

*Essay question: Select one aspect of the effects of solar flares on Earth and discuss it in depth in relation to one level of reasoning.*

**First sentence:** Solar flares affect climate conditions on Earth in various ways.

---

To provide you with a different question and a similar approach in answering it, here is another example:

**Example 12-5** **First sentence**

*Essay question: How do sperm cells of sea urchins and starfish differ?*

**First sentence:** The sperm cells of sea urchins and starfish differ in several main ways.

A different approach in writing the first sentence of the answer to an essay question is to pick a temporary working title for your essay. Your essay question may not ask for one, but a working title can keep you focused and ensure that you answer the right question and that you answer the entire question. The working title could be the topic, a category term, or the point of view that you intend to use. After you have come up with a working title, turn it into your topic statement by writing it as a complete sentence (and omit the working title from your essay).

**Example 12-6** **Working title**

*Essay question: How do sperm cells of sea urchins and starfish differ?*

**Working title:** "Sperm cells of marine invertebrates"

**First sentence:** The sperm cells of marine invertebrates differ in morphology, physiology, spawning habitat, and fertilization strategy.

## ➤ Logically organize and present your ideas

To compose the individual paragraphs of the body of your essay, use the bullet points you brainstormed and organized in your outline (see also Chapter 3). Begin each paragraph with a topic sentence; then, support the topic sentence with reasons and/or examples. Fill in the details, facts, and ideas necessary to support the points you have developed in your outline. Create good flow by connecting sentences either through placement of information by considering word location (see Chapters 2 and 3), repetition of key terms (see Chapter 3), or use of transitions (such as *first, second, next, finally, on the other hand, consequently, furthermore, in conclusion*; see Chapter 3).

Ensure that the information within your paragraphs is logically organized. Make your answer as specific as possible, and write only what you are asked to address. Stay within the word limit if one has been set. Also ensure that you structure the body of the essay in the same order as the topics listed in your introduction. Consider word location, and use signals and transitions to direct the reader through your points (see Chapters 2 and 3).

Give specific examples and details where possible. Unless you are explicitly asked to state your opinion in the question, do not write about your opinion or how you "feel." The instructor will be evaluating your understanding of the material, not your opinion or feelings.

Provide a clear conclusion to your essay. That is, end your essay with one or two sentences that summarize the main points of your answer. A well-written conclusion can be achieved by restating your central idea and indicating why it is important.

A short essay answer may look as follows:

---

**Example 12-7** **Short essay answer**

**Essay question:** *How do sperm cells of sea urchins and starfish differ?*

**Working title:** "Sperm cells of marine invertebrates"

First sentence

Body

The sperm cells of marine invertebrates differ in morphology, physiology, spawning habitat and fertilization strategy. For example, most sea urchin sperm cells are blunt ended at the acrosomal tip and contain only a small amount of F-actin prior to the acrosome reaction.

Their chemoattractant is resact, a 14-residue peptide. In contrast to sea urchin sperm cells, the sperm cells of starfish are more slender, have a much longer acrosomal tip with large amounts of F-actin, and have preformed acrosomal filaments. Their chemoattractant is startrak, a 13-kDa, heat-stable protein. In addition, whereas the sperm cells of sea urchins are rather species-specific, those of starfish are not.

Concluding sentence (optional)

The patterns of variation in sperm morphology and physiology may explain the mechanisms by which species recognition evolves.

---

In the following example of a longer, well-written essay answer, note the detail in the body of the essay, the flow, and the overall logical structure. The essay also includes a good opening sentence as well as a closing paragraph. The author does not include any opinion—simply facts.

---

**Example 12-8** **Long essay answer**

*Essay question: Describe the theory for the origin of life on Earth and how it is supported by evidence from the following areas:*

- *Origin of Earth*
- *The planet's early conditions*
- *Fossils, first living things, and conditions for life to flourish*
- *Molecular biology*

**Working title:** "Theory and evidence for the origin of Life on Earth"

First/introductory sentence and background

The theory for the origin of life is supported by several pieces of evidence, including the origin of the planet, its early conditions, fossil record, and molecular data. The Earth's formation is linked to the beginning of the Universe, for which the current theory is the Big Bang. This theory is supported by the Doppler effect because when objects are traveling away from a person, the wavelength of light lengthens, giving objects a redder appearance. Using spectrum analysis, scientists can tell what elements are within stars. When comparing the elements found in our sun to those of other stars, the same elements are observed. However, they are slightly reddened because they are moving away from us, giving them longer wavelengths. This observation suggests that the universe is expanding, which indicates that it had a beginning, as hypothesized in the Big Bang theory.

Body—detail 1

The theory for the formation of Earth is the dust cloud theory, which states that Earth formed from gaseous clouds that were disk-shaped. All the other things in the solar system formed similarly. Most of the hydrogen and helium in the solar system ended up in the sun, whereas most of the heavier elements formed the planets, including the Earth. In its infancy, Earth had little to no crust and an atmosphere of hydrogen and helium. This stage would have lasted no more than 100 million years. Earth's second atmosphere was composed of the gases volcanoes emitted: $CO_2$, $H_2S$, $H_2O$, ammonia, etc. Then, life began to form—probably in Earth's oceans—and eventually became photosynthetic. Photosynthesis caused oxygen to be produced.

Body—detail 2

Earth's third atmosphere, high in oxygen levels, supported aerobic life, which in turn evolved to become bigger and more complex. An ideal temperature for life existed because of the internal pressure, heat and radioactive decay, and the friction produced in the initial formation of Earth. The conditions on the early Earth would have made life itself almost inevitable. The early Earth, which had many volcanic gases as well as lightning, would have formed organic compounds. This has been repeated in experiments today. The oceans would have become an organic soup where protobionts would have eventually formed. Some of the compounds in the "soup" included microspheres and RNA, the latter of which has catalytic as well as informational properties. Furthermore, there was abundant energy and partial isolation, which would have allowed DNA to form. There were replicators, and they had the potential to mutate.

| | |
|---|---|
| Body—detail 3 | The theory of life is also supported because the first prokaryotes were very simple organisms that could have formed from this "soup." Fossils show that bacteria did exist about 3.8 billion years ago. The Greenstone belt and biogenic graphite in western Greenland are the oldest rocks that suggest life on Earth. Bacteria were extremely simple and used carbon dioxide as a carbon source, but soon began to evolve into more complex forms. |
| Body—detail 4 | Molecular biology also supports the theory of life on Earth because it shows that biological molecules have—like organisms—themselves evolved. For example, RNA, an information-storing system, can form itself into a ribozyme under certain conditions. This ribozyme in turn acts as a catalyst for RNA formation. Catalytic functions were eventually taken over by enzymes because they were more efficient. Therefore, a molecular evolution took place. |
| Concluding paragraph | All together, these pieces of evidence provide us with information on the formation and origin of Earth. They are the basis for the theory for the origin of life on the planet. |

*(With permission from Nikolas Franceschi Hofmann)*

### ➤ Proofread your answer

After you are done writing your answer, read over it and check that all the main ideas have been included. An essay answer also tests your writing ability. Thus, be sure to review and proofread your answer for careless mistakes, misspelled or illegible words, and grammatical errors.

### ➤ Be aware of additional strategies to answer an essay question

Additional strategies may include:

- Paying attention to the rubric if provided by the instructor; use it as a guide to highlight specific areas on which to focus your answer.
- Using relevant technical terminology that you learned from your courses to answer the question.
- Supporting your answer with evidence and/or examples from class.
- Providing a drawing or diagram that can assist you in your answer; you will need to explain what the drawing is in writing, however.
- Making relevant and logical connections clear—for example, for cause-and-effect relationships or drawings with labels.
- Budgeting your time to ensure that you are not forced to rush through your essay.
- Leaving time to proofread for grammar, spelling errors, omissions, etc. Too many errors will be credibility killers, but do not avoid words because you cannot spell them.

- Qualifying your answers when in doubt. It is better to write "about 500" than "485" when you are unsure about the exact number.
- Avoiding excess information.
- Writing down your ideas for potential credit in case you do not know the exact answer or run out of time when you are writing an answer; do not leave an answer blank.
- Writing legibly so your instructor can read what you have written and you do not lose points needlessly.

### ➤ Take advantage of campus resources

Seek out resources available on your campus to help you succeed. Every campus is different, but many institutions have writing centers, tutoring centers, disability offices, and learning centers, which can help students in small group settings or individually. You can also find out the office hours for your instructor and attend them prepared with specific questions. You will find that instructors are more than willing to explain in more detail, point you in the direction of additional help, and even hint at potential test questions. Know that these resources are not just for struggling students, but are for all students; even the best students take advantage of them.

## 12.5 TIME MANAGEMENT TIPS

### ➤ Manage your time efficiently

Be aware of the time, especially if there is more than one essay question you are asked to answer. Read over the exam and instructions for each question carefully. Then, decide which question(s) you will answer first. Start with the one you know most about and the one you can answer the quickest.

As a guideline, one- to two-sentence answers (20 to 40 words) will take generally 2 to 5 minutes to write and will leave you no time for details on a major point; a short-paragraph answer (50 to 100 words) will take 15 to 20 minutes to write, allowing time for one detail on a key idea; a longer or multiple paragraph answer (100 to 200 words) will take 20 to 30 minutes to compose and will give you time for two details on a key idea. Even longer essay answers (200 to 600 words) will require 30 to 60 minutes to compose; instructors may consider giving these as take-home essays rather than in-class tasks if time is limited during an exam.

## 12.6 CHECKLIST

Use the following as a checklist after you have written your answer.

**Checklist for Essay Tests**

- ☐ Will the reader get an overview of your answer from reading the first sentence of your answer?
- ☐ Did you describe the best ideas first?
- ☐ Did you support your major ideas with examples and facts?

- ☐ Did you answer the question fully?
- ☐ Did you include only directly relevant material?
- ☐ Did you end with a conclusion?
- ☐ Did you proofread your essay?

## SUMMARY

1. To take good notes:
   - Write notes down on paper.
   - Use shorthand, abbreviations, and symbols.
   - Organize and structure your notes as you go.
   - Draw pictures.
   - Pay attention to content.
   - Go over your notes right after the lecture.
2. Study your notes.
3. Prepare for an essay exam—study and understand the material.
4. To answer an essay question:
   - Read and understand the question.
   - Plan your response: Jot down ideas/brainstorm and make an outline.
   - Write the first sentence as an overview sentence.
   - Logically organize and present your ideas.
   - Proofread your answer.
5. Be aware of additional strategies to answer an essay question.
6. Take advantage of campus resources.
7. Manage your time efficiently.

# PART IV

# Advanced Scientific Documents and Presentations

CHAPTER 13

# Oral Presentations

THIS CHAPTER DISCUSSES:

- How to prepare for a presentation
- Content and organization of a scientific talk
- Visual aids
- How to present your talk
- How to deal with questions and answers

## 13.1 FORMAT OF A SCIENTIFIC TALK

### ➤ Organize your slides

The content of a scientific talk is similar to that of a research article with an introduction, purpose, experimental design, results, and discussion/conclusion. Unlike a research article, however, an oral presentation needs to be structured and worded differently from a written document.

Whether you are presenting a short, 10-minute conference talk or a full-hour seminar, a good overall format to follow for your talk is the following:

| Title slide (optional) | |
|---|---|
| **First slide:** | **Overview of talk (optional for short talk)** |
| **Next slides:** | **Introduction/Background and purpose of study** |
| **Subsequent slides:** | **Findings (combined with general approach)** |
| **Final slide:** | **Conclusions and main supporting points** |
| Credit slide (optional) | Acknowledgment of those who worked with you or financed your research. |

### ➤ Identify the purpose/question and the take-home message of the study

Similar to that of a research paper, the central theme of your talk needs to revolve a specific question, that is, the purpose of your study. Identify this purpose/question clearly, and present it right after your introductory slide. (See also Chapter 7, Section 7.6, on abstracts.) Next, pinpoint the take-home message of the talk. Together with the purpose/question, the take-home message will serve as the filter for what to include in the talk. All other slides and data of your talk will be selected based on the question and take-home message, which should be presented at the end of your talk.

### ➤ Pick out the most important figure for your core slide

When you are preparing your talk, you need to ensure that you address the most important information, no matter what the time limit. To help you select your most important figure or core slide, imagine you were given only 2 minutes to present your work. The core slide will be the single slide you would need to include to make your talk compelling. The core slide needs to be clear, uncluttered, and immediately understandable. It should contain your most important figure, be highly visual, and be the slide(s) that everyone remembers when they leave.

### ➤ Know your audience

It is critical that you know your audience and the overall purpose of your talk before you prepare your slides. Depending on your target audience and the type of talk, different emphasis may have to be placed on different sections of the presentation. For example, if your talk is geared toward a nonscientific audience, you may have to include more introductory material and slides, because it will be particularly important to provide sufficient background information to avoid losing your audience in the first few minutes. For a scientific audience, on the other hand, introductory slides may not be as critical as more details on results, control experiments, and comparisons and contrasts to findings of others.

### ➤ Prepare an overview slide for longer presentations

Start your presentation by telling the audience what you will speak about, mention how you will present your findings and that you will summarize all before you close. Then, move on to provide general background information on your topic.

If you are preparing a longer presentation, you may want to consider using an overview slide at the very beginning to lay out the different parts you will be covering. Use the same items as slide titles later in the talk. Follow the overview slide with an introduction of background material to provide context for your talk.

After the background, explain the purpose of your studies and discuss your results. The main emphasis should be on your key findings. The experimental approach is often only briefly alluded to verbally and may not require a slide. Providing an overview of the experimental approach is sufficient; exact details such as amounts of reagents are usually not given in a presentation. However, if the experimental setup is complicated or if you are introducing new or complex methodology, you may need to explain the experimental approach in more detail to your audience. In these cases, including additional slides on your approach is appropriate and necessary.

Most presentations end with a concluding slide, which summarizes and generalizes your main findings. Depending on your audience, a concluding slide may also point out next steps or give credit to colleagues and collaborators.

Shorter talks are usually more difficult to prepare and to present than longer talks. For shorter talks, you need to be extremely selective about what information to present. For such talks, concentrate on the overall, most important information for each of the sections. Generally, this means that you will need to reduce the number of background slides and focus on one to three main findings and their interpretations.

## 13.2 PREPARING FOR A TALK

### ➤ Prepare your talk and visual aids well ahead of time

The two most important points in becoming a good presenter are to be prepared and to practice. Know your audience and their level of expertise. Design your talk accordingly in terms of its direction and necessary background information. Put your slides together well ahead of the talk. Preparing visual aids well ahead of the presentation will give you time to check, replace, or improve them.

### ➤ Make slides look attractive, but keep them simple

Visual aids should be comprehensible on their own. They must be clear, legible, and easy to understand. They should also be visually pleasing, professional, and free of spelling errors.

For scientific presentations, most of the time you need to use a conference-style presentation for your slides. The intention in this type of presentation is to either inform or persuade your audience. Therefore, provide sufficient information in text slides as well as with figures and tables to present an overview of your work or specific topic without an overload of information, text, or figures.

Any graphical layout must immediately convey what the slide is about, must contain relevant details, and must be free of distractions. Note that although slide layouts with large photos and no or only sparse text are favored by some schools or organizations, these are best for keynote addresses, courtrooms, and sales meetings—not for scientific conferences.

### ➤ Know what programs are available to create slides

For the most part, Microsoft PowerPoint, Apple Keynote, SlideRocket, Google Slides, Adobe Presenter, and so forth can be used to design your slides for a scientific presentation. All of these presentation tools follow the traditional outline, linear slide presentation path. Another alternative is Prezi, which uses a zoom-in/zoom-out, creative, nonlinear format to presenting your ideas. Generally, similar zoom-in and nonlinear style can also be achieved with the more traditional programs mentioned earlier, although information on how to do so is not as readily obvious in these programs. If you opt for a zoom-in/zoom-out presentation, ensure that you do not make your audience dizzy with too many (or too quick) transitions.

### ➤ Pay attention to color and design

To create visually attractive slides, avoid bright colors, nonstandard colors, and red/green (or blue/orange) color contrasts, as some people are color blind to these. Individual slides should also be recognizable as being part of a set (same colors, style of fonts, and emphasis techniques). In addition, ensure slide that titles are always in the same place on each slide. Do not use wildly animated slide transitions or similarly animated bullet points. Ensure that punctuation is consistent throughout the presentation.

### ➤ Pay attention to font

Avoid nonstandard fonts, graphs, and abbreviations. Ensure that the minimum font size is larger than or equal to 20 points so lettering can be read by the audience during the presentation. Use a sans serif font such as Arial (without the projections at the end of letters) or serif such as Times New Roman (with projections) on your slides. However, for figures, prefer a sans serif font type—this is usually easier to distinguish from data in graphs. Look for high contrast between background and writing or figures. Choose dark text against a light background, or vice versa. Black text on white background or white or yellow writing on dark blue background, for example, is commonly used and easy to read. If you need to emphasize text, put the most important information in a larger print size or into italics, add an arrow, or use a different color.

### ➤ Limit the amount of text

Not only do your slides need to be visually attractive, they also should contain informative headings and cover all key points, yet be simple. Most people can only absorb a limited amount of information at once. Therefore, to ensure that you are not losing your audience, do not clutter your slides with too much text. In an average presentation, you will spend about 1 to 2 minutes on each slide and will have to cover all the information that is contained in it.

For text slides, write clear but brief bullet points to provide an overview, and fill in the rest with your voice. If your slides contain too much text, your audience will concentrate on reading the text and not listen to what you have to say, or they will be listening to you and not pay attention to what is on the slide. Neither case is what you as the presenter really desire.

As a rough guideline, use the 5x5 rule:

- About five words per bullet point
- Five bullet points per slide

Do not use more than about 50 words per slide. An exception to this guideline is when you prepare slides that you plan on sending out instead of presenting directly to an audience.

Following is an example of a slide that contains too much text:

**Example 13-1** **Text slide**

Overview of the Yale School of Medicine

- Founded in 1810, the Yale School of Medicine is a world-renowned center for biomedical research, education, and advanced health care.
- The school is viewed internationally as a leader in biological and medical research.
- The Yale School of Medicine has over 2,300 faculty members and consists of 10 basic science departments and 18 clinical departments.
- The School of Medicine consistently ranks among the handful of leading recipients of research funding from the National Institutes of Health and other organizations supporting the biomedical sciences.

To improve a text-heavy slide such as the one shown, you need to decide which bullet points can be omitted and what information in the remaining bullet points is really important. Then, list the important information by itself in a bullet point and omit the rest. A possible revision follows:

**Revised Example 13-1a** **Text slide**

Yale School of Medicine Overview

- Founded in 1810
- Leader in biomedical research
- Over 2,300 faculty members
- 10 basic science departments
- 18 clinical departments
- Top biomedical research funding

In the revised version, the bullet points have been reduced to their main piece of information and are visually distributed better on the slide, all of which is preferred by the audience.

If we further revise this example by adding a different background as well as a picture and using different styled and shaded text for the heading, we get a much more visually attractive slide:

**Revised Example 13-1b** **Text slide with background and graphic**

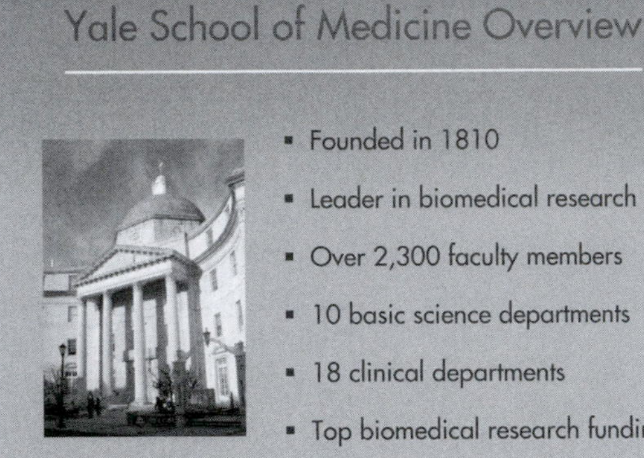


### ➤ Give figures and tables a title but no legend or caption

On slides, figures and tables should have a title. Present only figures or tables with a heading large enough to be visible from the back of the room. Do not include a figure legend or caption on slides. Instead, fill in this information verbally.

### ➤ Think graphically

Know that readers prefer visually well-designed slides rather than text-heavy slides. They also prefer graphs rather than tables or text, particularly bar graphs because their message can be quickly understood. However, do not simply transfer published figures or tables onto your slides. Their lettering is often too small, they do not have a heading, and more often than not, printed figures and tables contain additional information that is not needed for your presentation. Instead, recreate and simplify graphs and charts. Ensure also that axis labels and keys are clearly visible.

### ➤ Do not clutter figures and tables

Ensure that your figures and tables are not cluttered or too busy. For line graphs, display not more than three to four curves per graph. Similarly, in bar graphs, keep the number of bars to a minimum. In your figures, symbols and shading must be easy to tell apart. Provide a key for figures if needed. An example of a slide containing a graph is shown next. This graph contains just enough information for the presenter to walk the reader through all the information in 30 to 60 seconds and for the reader to digest this information in the same time frame. Note that on this slide, the title for the graph is at the same time the title for the slide. It is possible to compose such titles if your slide contains a single figure.

**Example 13-2a Slide with bar graph**

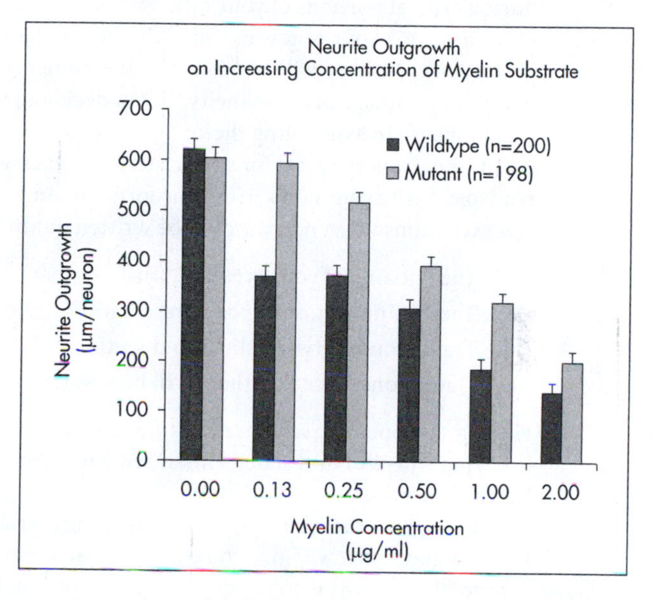


*(With permission from Betty Liu, modified)*

If you have to use a table, aim for a maximum of four columns and seven to eight rows, including the title and column headings. Write all text horizontally. The next example shows a slide containing a well-constructed table.

**Example 13-2b Slide with table**

**Thermal induction of cytokinin overproduction**

| | **Concentration of cytokinin in medium (nM)** | |
|---|---|---|
| | **15°C** | **25°C** |
| **Mutant and Cytokinin** | | |
| **thiA1 + iP** | 0.16 | 0.23 |
| **thiA1 + [9R]iP** | 0.09 | 0.16 |
| **OveST25 + iP** | 6.34 | 60.0 |
| **OveST25 + [9R]iP** | 0.54 | 9.51 |

### ➤ Prepare notes in large print, and memorize the opening sentences

Most people find it helpful to prepare notes for their presentation. You may not need them when you are presenting but might want to use them to practice your talk. For most people, notes also provide the security of knowing what to fall back on when you are too nervous to remember anything during critical portions of your talk.

To record your notes, use index cards or the speaker's notes field on PowerPoint. Neither will be visible to the audience and will allow you to maintain an image of spontaneity. If you decide to use note cards, number them for ease in assembling them.

Use extra-large print for your notes if necessary—they should be easily readable. Write your notes in outline form, not in full sentences. There are a few exceptions when notes should be written out in full:

- The opening of your presentation
- The closing section of your presentation
- Transitions (between slides and sections)
- Quotations (note that there will be few, if any, in scientific talks)

Having the opening sentences available in full will give you comfort in knowing where to find statements when you need them, and when you are the most nervous.

To deliver the opening sentences firmly and accurately, memorize them. A good start will give you confidence to continue and allow nervousness to abate. As you conclude your presentation, it is also comforting to know that you have your conclusion written out, especially to ensure that you get the wording correct. Transitions that easily evade you during your talk should also be written out in full to serve as bridges and references.

### ➤ Practice, practice, practice

For the best possible presentation, it is of utmost importance that you practice and practice and practice. Without practicing, you will not know whether you stay within the given time limit for your talk. Nor will you know if your flow of words is smooth or if your voice has the right pitch. Practicing will make you realize at least some of these potential problem areas. Consider videotaping yourself, or ask someone else to do this for you. Review the video, and note any areas that may need improvement. In addition, take advantage of opportunities to give practice presentations.

## 13.3 DELIVERY OF A TALK

### ➤ Arrive early, and dress appropriately

When you have to present a talk, arrive early to allow sufficient time to familiarize yourself with the surroundings and equipment and to load your presentation. Ensure that your presentation file is compatible with the computer connected to the LCD projector, especially if you deliver your presentation on a flash drive or CD. As a back-up, consider bringing your own computer, store your presentation on an online cloud, or send a PDF copy of the file to yourself or the organizer ahead of time. Alternatively, come prepared with several copies of a printed-out version of your slides, or with notes pages, which you may ask to have photocopied and handed out if all else fails.

Dress appropriately to convey professionalism. That usually means business casual to formal, depending on the audience and purpose (i.e., a shirt, nice pants, slacks or skirt [not too short!] as well as nice shoes). Avoid chewing gum, flip flops, smelly foods, and unkempt hair.

### ➤ Use spoken English

During your talk, use simple words even if you have to apply technical terms. Do not speak the way you write in scientific English. Use spoken language instead.

If you use notes during your presentation, do not read them word for word. Also, do not use the bullet points on your slides as your notes. These bullet points are intended for your audience, not for you as the speaker. Therefore, it is important that you prepare separate notes for yourself. When you use notes, speak and act normally. Being yourself is the most valuable asset of a speaker. Above all, be positive and enthusiastic about your topic.

### ➤ Stick to the time limit, and speak slowly

It is customary to assign a limit to the length of a presentation. Adhere to this time limit. One of the most annoying things for any audience is a talk that goes overtime. To help you keep track of time, place a timer on the lectern. On average, you will need 1 to 2 minutes per slide. If you find that you need to leave out some important slides to stay within the time limit, do so. Do not speed through your talk. You run the risk of losing your audience if you present your material too fast. Instead, pace yourself. Remember: Although you are familiar with the material and have gone over it many times, your audience will hear it for the first time, so give them a chance to digest it. A good talk requires speech that is slower and clearer than normal conversation.

### ➤ Make sure you can be heard by the entire audience

Do not speak too softly. Soft speech signals that the speaker is uncertain. However, do not blast the audience out of its seat by the volume of your delivery either. In addition, pay attention to the pitch of your voice. Speaking in deeper, fuller tones makes your voice more pleasant to listen to.

### ➤ Avoid distracting sounds

If you use too many distracting sounds such as "hm" and "uh," "ok," and "so," you will appear unprepared and unsure of what will come next. These sounds, if used in excess, can get very distracting for your audience. Avoid these distractions.

### ➤ Stay within the presenter's triangle

The best position for a presenter is on the left or right side of the room or screen (Example 13-3). Either of these positions will ensure that you do not block the projection of your slides. It will also ensure that you do not block the view of the audience.

**Example 13-3 Ideal presenter's location**

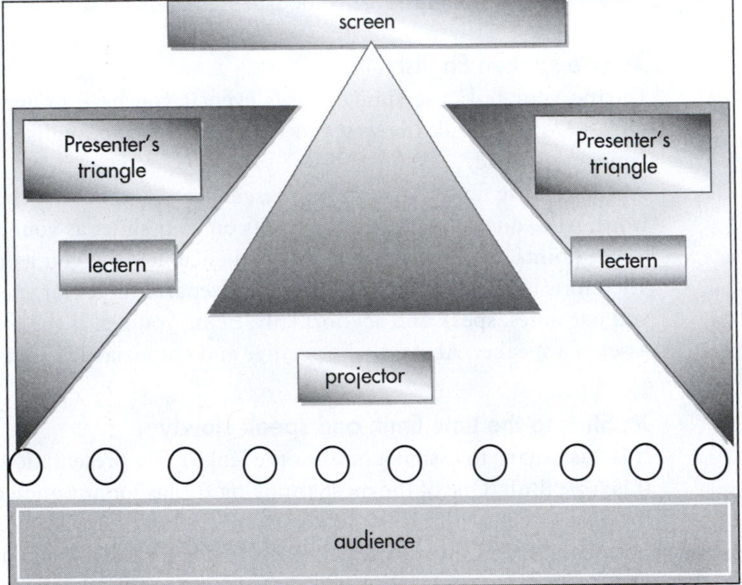


### ➤ Keep eye contact, and face the audience

In any presentation, confidence is the key. The most powerful way to appear confident is to look directly at the audience. Thus, be very conscious of head and eye movements. Eye contact with the audience is essential. The best way to keep your eyes on the audience is by keeping the front of your body facing the audience as much as possible. Look at individuals in the audience, but do not exclude sections of listeners.

### ➤ Use gestures

Use gestures to reinforce and complement your talk. When you are not using your hands and arms, let them hang naturally, but do not stand rigidly. Use good posture. Do not stick your hands into trouser pockets or fold your arms or hands. Also, do not use your hands to straighten your clothing, rub your nose, explore your ears, smooth down your hair, or play with keys, bracelets, and so forth.

### ➤ Be conscious of body movement

Your feet should neither be seen nor heard. They should be securely on the floor. Do not teeter back and forth, and do not pace. Stay at the lectern unless you need to point out data on a slide or exhibit. When you show slides, turn halfway toward the screen rather than turning your back on the audience. Face the audience again after pointing out relevant parts on a slide.

If you are using a laser pointer, learn how to use it correctly and sparingly. Above all, learn where the "Off" button is. If your hands tend to shake while you are pointing, consider supporting the pointing hand with your other hand, or resting it on the lectern. Make a circling motion with the laser pointer on any word or figure you want to emphasize. Do not try to hold it still on one point.

### ➤ Explain everything on your slides

Simultaneously show on your slides the information you are providing verbally. Ensure also that you explain everything that is on a slide. Do not skip over information. When you present a figure or table, explain why you are showing the figure or table. Then, walk your audience through them in their entirety. For a graph, explain each axis first, and then explain the data and the results they represent. For a table, describe the row and column headings, and then list the key findings in the table field. Explain what the findings mean in the overall context of your presentation.

### ➤ Make your talk flow well

To make your presentation flow smoothly from one idea to the next, build in transitions between slides and around the bigger parts of your presentation. Without transitions, your talk will appear choppy, and sound just like a list of facts. As transitions are often forgotten, consider writing them down in your notes as reminders. Examples would be "An example of such a phenotype can be seen in the next slide" or "Our results show that . . ." or "In conclusion, . . ."

### ➤ Signal the end

When you conclude your presentation, signal the ending. Say "To conclude, . . ." or "In my final slide . . ." Then, summarize your findings and their meaning, briefly discuss the overall significance of your work, indicate next steps if appropriate, and acknowledge those who contributed to the work.

If your words and the tone of your voice do not make it clear that you have finished, just say "Thank you" and stop. If questions are to follow immediately, stay at the lectern. Otherwise, gather your items, and walk back to your seat.

## 13.4 QUESTION-AND-ANSWER PERIOD

### ➤ Ensure that you are in charge, but stay calm and polite

Often, a scientific presentation is followed by a brief question-and-answer period during which anyone in the audience can ask a question of the presenter. Many, if not most, presenters are nervous about this period, especially about being asked questions to which they might not know the answer. Do not get frazzled. You do not need to know all the facts. However, when asked about an opinion, you should be able to answer with such.

To deal with difficult questions:

**Do**

Be courteous in your answers at all times.
Repeat the question to ensure the entire audience has heard it clearly.
Admit when you do not know the answer.
Ask the questioner to rephrase the question.
Direct the answer to the entire audience.
Ask the questioner to talk to you after the session.

**Don't**

Do not give your audience an opportunity to interrupt your presentation.
Do not make up an answer if you do not know.
Do not maintain eye contact with the questioner when you give your answer.
Do not argue with a questioner.

## 13.5 USEFUL RESOURCES

For the top 20 tips in using PowerPoint, see Appendix D. For more suggestions and advice on how to create, prepare, and present a scientific oral presentation, consider also the following websites (last accessed February 2018):

- **http://www.nature.com/scitable/topicpage/oral-presentation-structure-13900387**—excellent resource that discusses the structure of an oral presentation
- **http://www.nature.com/scitable/ebooks/english-communication-for-scientists-14053993/giving-oral-presentations-14239332**—another site that discusses scientific talks in good detail
- **http://srp.lsu.edu/ResourcesforStudents/item50702.pdf**—a step-by-step guide on how to prepare and deliver an oral presentation
- **http://www.northwestern.edu/climb/resources/oral-communication-skills/creating-a-presentation.html**—useful guide on how to prepare an oral presentation
- **https://www.youtube.com/watch?v=LzIJFD-ddoI**—good video explaining how to prepare on oral research presentation

For useful tutorials of PowerPoint, see (last accessed February 2018):

- **http://www.lynda.com/landing/msofficetutorials.aspx**
- **http://www.actden.com/pp/**
- **https://www.udemy.com/courses/search/?q=Powerpoint**

## 13.6 CHECKLIST

### Overall

- ☐ Did you practice your talk several times?
- ☐ Is your talk within the time limit?
- ☐ Are you aware of distracting sounds or habits you show when presenting?

### Preparation

- ☐ Did you find out who your audience will be?
- ☐ Did you prepare notes?
- ☐ Did you write down:
    - ☐ Your opening statement?
    - ☐ Important transitions?
    - ☐ Your concluding remarks?

### Slides

- ☐ Do you have an overview slide?
- ☐ Is your presentation logically organized?
- ☐ Are your slides informative?
- ☐ Do you have a summary slide?
- ☐ Is each slide logically organized and uncluttered?
- ☐ When preparing your slides, did you concentrate on the main points in each portion of your talk?
- ☐ Do all figures and tables have a title?
- ☐ Did you use visuals where possible rather than text?
- ☐ Is text used sparingly but informatively?
- ☐ Is the font large enough?
- ☐ Are slides/figures/tables kept simple?
- ☐ Are slides attractive? Is color used well?
- ☐ Did you proofread your text?

## SUMMARY

### ORAL PRESENTATION GUIDELINES

1. Organize your slides:

| | |
|---|---|
| Title slide (optional): | |
| First slide: | Overview of talk |
| Next slides: | Introduction/Background and purpose of study |
| Subsequent slides: | Findings (combined with general approach) |
| Final slide: | Conclusions and main supporting points |
| Credit slide (optional): | Acknowledgment of those who have worked with you or financed your research |

2. Identify the purpose/question and the take-home message of the study.
3. Pick out the most important figure for your core slide.
4. Know your audience.
5. Prepare your talk and visual aids well ahead of time.
6. Prepare an overview slide for a longer presentation.
7. Make slides look attractive, but keep them simple.
8. Know what programs are available to create slides.
9. Pay attention to color and design.
10. Pay attention to font.
11. Limit the amount of text.
12. Give figures and tables a title but no legend or caption.
13. Think graphically.
14. Do not clutter figures and tables
15. Prepare notes in large print, and memorize the opening sentences.
16. Practice, practice, practice.
17. Arrive early and dress appropriately.
18. Use spoken English.
19. Stick to the time limit, and speak slowly.
20. Make sure you can be heard by the entire audience.
21. Avoid distracting sounds.
22. Stay within the presenter's triangle.
23. Keep eye contact, and face the audience.
24. Use gestures.
25. Be conscious of body movement.
26. Explain everything on your slides.
27. Make your talk flow well.
28. Signal the end.
29. Ensure that you are in charge, but stay calm and polite.

# PROBLEMS

## PROBLEM 13-1

**Assess the following slide. Explain why this is not a good slide.**

**TABLE 2**

List of Proteins Renatured by the Sparse Matrix Approach and Their Optimum Renaturation Conditions

| Protein | Molecular mass (Da) | Structure | Standard assay buffer | Optimum renaturation buffer system | Max activity recovered (percentage of initial value) |
|---|---|---|---|---|---|
| BAP | 80,000 (29) | Homodimer $Zn^{2+}$, $Mg^{2+}$ cofactor | 100 mM NaCl, 5 mM $MgCl_2$, 100 mM Tris, pH 9.5 | 0.2 M Na acetate, 0.1 M Tris–HCl, pH 8.5, 30% (w/v) PEG 4000 | 138 |
| HRP | 40,000 (30) | Monomer, heme group, $Ca^{2+}$, carbohydrate | 100 mM $CH_3COONa$, pH 4.2 | 0.2 M Mg acetate, 0.1 M Na cacodylate, pH 6, 30% (v/v) 2-methyl-2,4-pentanediol | 33 |
| β-gal | 540,000 (31) | Tetramer, with independent active sites (32) | Z-buffer (see Materials and Methods) | 30% (v/v) PEG 1500 | 81 |
| Lysozyme | 14,388 (33) | Monomer | 0.1 M K phosphate buffer, pH 7.0 | 0.2 M Mg acetate, 0.1 M Na cacodylate, pH 6, 30% (v/v) 2-methyl-2,4-pentanediol | 333 |
| Sperm bindin | 24,000 (23) | Unknown | Seawater | 0.2 M Na citrate, 0.1 M Na Hepes, pH 7.5, 30% (v/v) 2-methyl-2,4-pentanediol | 100 |
| Recombinant bindin | 24,000 (34) | Unknown | Seawater | 0.1 M Na Hepes, pH 7.5, 1.6 M Na, K phosphate | 100 |
| Trypsin | 23,800 (35) | Monomer (α) dimer (β) | 10 mM Tris, pH 8.0 | 0.2 M Mg chloride, 0.1 M Na Hepes, pH 7.5, 30% (w/v) PEG 400 | 11 |
| Acetylcholinesterase | 260,000 (36) | Aggregates, monomer | 0.2 M Na phosphate buffer, pH 7.0 | 0.1 M Na Hepes, pH 7.5, 0.8 M Na phosphate, 0.8 M K phosphate | 9 |

*(With permission from Elsevier)*

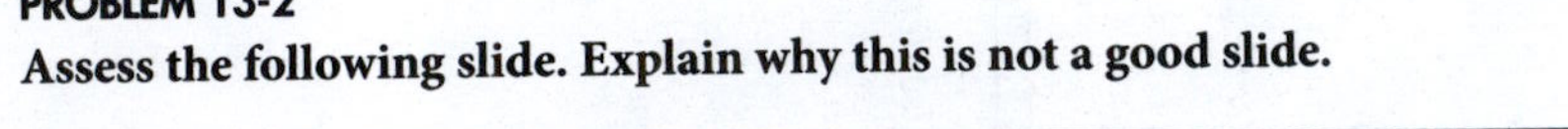

**PROBLEM 13-2**

**Assess the following slide. Explain why this is not a good slide.**

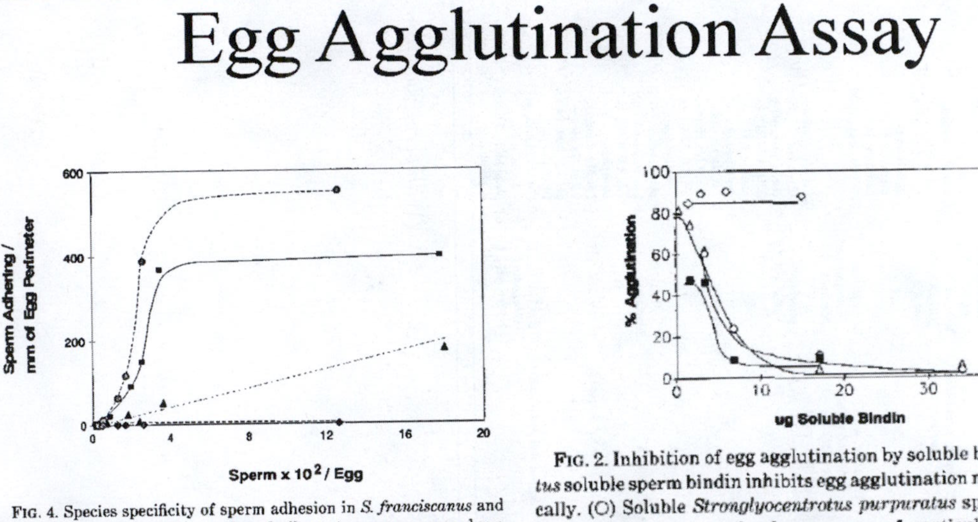


FIG. 4. Species specificity of sperm adhesion in *S. franciscanus* and *S. purpuratus* gametes. The number of adherent sperm was scored as a function of sperm concentration for all possible combinations of *S. franciscanus* and *S. purpuratus* gametes. ●, *S. purpuratus* sperm × *S. purpuratus* eggs; ◆, *S. purpuratus* sperm × *S. franciscanus* eggs; ▲, *S. franciscanus* sperm × *S. purpuratus* eggs; ■, *S. franciscanus* sperm × *S. franciscanus* eggs. Significant numbers of *S. franciscanus* sperm adhere to the surface of *S. purpuratus* eggs. The number of adherent sperm is normalized to account for the larger surface area of *S. franciscanus* eggs.

FIG. 2. Inhibition of egg agglutination by soluble bindin. *S. purpuratus* soluble sperm bindin inhibits egg agglutination non-species specifically. (○) Soluble *Strongylocentrotus purpuratus* sperm bindin added to *S. purpuratus* eggs in the presence of particulate *S. purpuratus* bindin; (△) soluble *S. purpuratus* sperm bindin added to *S. purpuratus* eggs in the presence of particulate *S. franciscanus* bindin; (■) soluble *S. purpuratus* sperm bindin added to *S. franciscanus* eggs in the presence of particulate *S. franciscanus* bindin; (◇) soluble synthetic peptide corresponding to residues 69–130 of *S. purpuratus* bindin added to *S. purpuratus* eggs in the presence of particulate *S. purpuratus* bindin (see text).

*(With permission from Elsevier)*

**PROBLEM 13-3**

**Explain why the following statements are not good choices for an oral presentation:**

1. "Thank you for listening to my talk. I hope it was not too confusing or boring."
2. "On this slide, please focus only on part D, and ignore parts A, B, and C."
3. "This finding is in agreement with that of a previously published result reported by Lopez et al. in 2001, where it was shown that emergence of seedlings is temperature- and humidity-dependent."
4. "It was determined that frogs can hibernate under water for up to 6 months."

**PROBLEM 13-4**

**The following slide is poorly worded and text-heavy. Follow the 5x5 rule and reduce the amount of text to maximum five words per bullet point.**

**2014 Focus Group**

Patients emphasized the value of:

- The welcome environment of the clinic
- Being able to communicate clearly in their native language
- The comprehensiveness of care offered during a visit
- Being seen by professional and caring clinical teams
- Feeling respected as human beings

CHAPTER 14

# Posters

THIS CHAPTER DISCUSSES:

- Format and organization of a poster
- Focus of a scientific poster
- How to prepare the different components of a poster
- How to prepare a conference abstract
- How to present a poster

A scientific poster is a visual presentation of your work that can serve as a concise communication tool for small groups of people. The poster should attract viewers, serve as an advertisement, and provide an overview of your work in your absence. Scientific posters differ from those in other discipline in that they require a specific format with certain scientific details.

## 14.1 COMPONENTS AND FORMAT OF A POSTER

### ➤ Design the poster around your research question

Include:

Title
Abstract (optional)
Introduction
Materials and Methods
Results
Conclusions
References
Acknowledgment (optional)

Overall, posters follow the standard scientific format in that they contain all the sections found in a research paper except for the Discussion. Posters include: Title, Abstract (optional), Introduction, Materials and Methods, Results, Conclusion, and References, and optional are Acknowledgments. However, posters do not contain all the details and information found in a manuscript. That is, their word count is much less. Instead, poster sections concentrate only on the main points of each section and present these mostly visually.

### ➤ Find visual ways to show your work—let the illustrations tell the story

Think of each part of the poster as a slide that you would show to an audience. As done for a slide presentation, try to keep text to a minimum. The illustrations need to tell the story.

Posters are typically presented to small groups of people interested in hearing more about your topic and work. The clearest poster presentation stems from the proper arrangement of information and from simplicity of design. Because the visual impact is most important for a poster, use illustrations, symbols, colors, and so forth wherever possible rather than text.

### ➤ Focus only on the main points in each section

Like papers, posters should be designed around your research question. During your poster presentation time, you can expand further on this research question, approach, and more. In addition, prepare a brief, distinct, and memorable take-home message. State implications and conclusions clearly, and gear them toward your audience.

### ➤ Aim for about 20% text, 40% graphics, and 40% blank space

Effective poster design is important. One of the first things you should do when you find out that you have to present a poster is to find out how much space you are allowed. Then, decide on your poster layout.

Sample horizontal poster layouts are shown in Figures 14.1a to 14.1c, but know that poster layouts and designs can vary widely. Some have a symmetrical layout while others do not. Some are horizontally laid out; others are vertical posters.

### ➤ Lay out the poster in a logical, easy-to-follow way

Although layouts can vary, viewers expect to find certain sections in specific places. The most important text sections (Abstract and Conclusion) are usually placed in the top left and bottom right corner, respectively, as readers read from left to right and top to bottom. The least desirable real estate on a poster is usually on the very bottom. Often, References, Acknowledgments, and logos are placed there. For clarity:

- Present the information in a sequence that is easy to follow. Consider numbering panels to help the audience follow the flow of the sections of the poster.

(a)

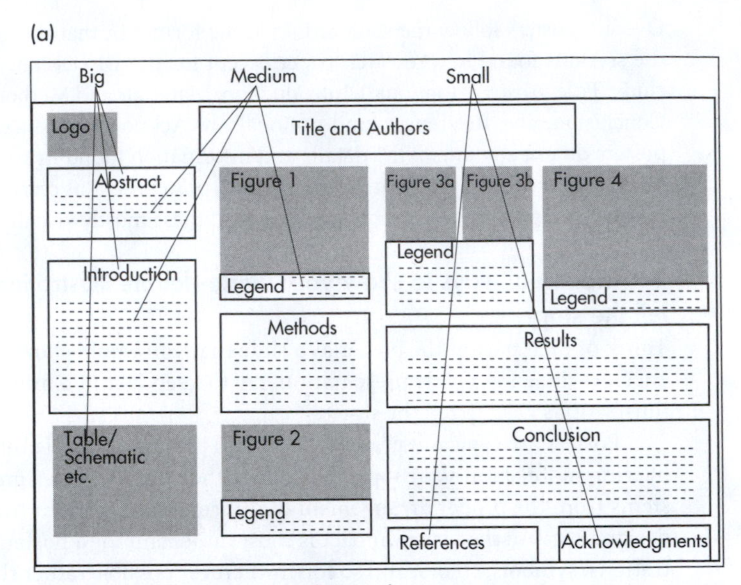


Figure 14.1a Sample horizontal poster layout, asymmetric.

(b)

Figure 14.1b Additional sample horizontal poster layouts.

- Arrange the material into columns—most horizontal posters allow for three to four columns, vertical posters usually contain two columns (see sample poster layouts in Figures 14.1a to 14.1c.)
- For maximum visual layout and impact, aim for about 20% text, 40% graphics, and 40% empty space as mentioned. Note that many posters contain much more text. Some even seem to display the pages from entire articles. Such posters are generally considered unattractive by viewers because most people do not want to read through all this text to get to the bottom line.
- Maintain a consistent style.
- Use bullets and numbers to break text visually and make it more readily available.
- Keep key terms consistent throughout.

(c)

Figure 14.1c Sample vertical poster layout.

### ➤ Design backgrounds well

The background design is important in the presentation of your data. For well-designed poster backgrounds, use white or muted colors. These colors are easiest to view for a long time and offer the best contrast for text, graphic, and photographs. Do not use more than two or three colors in the background, and do not use bright colors or background designs. All of these can be overbearing and distractive for your audience. Background design or a photo background can make distinguishing and reading both text and figures very difficult.

### ➤ Make fonts large enough to read

Because people are attracted to posters that have good graphics, a clear title, and few words, your main objective in preparing text for a poster presentation is to edit it down to very concise language. Ensure that illustrations are large enough to be clearly discernible from two yards away. Use the same font type throughout, preferably Arial, Times New Roman, or similar for ease of reading. As a general guideline:

Titles: 90 point, boldface
Subtitles: 72 point
Section headings: 32–36 point
Body text of main sections: 28–36 point
Other text (figure legends, references): 22–28 point

## 14.2 PREPARING A POSTER

Posters can easily be designed using computer programs such as InDesign, LaTeX, Illustrator, CorelDRAW, Microsoft PowerPoint, and more. Free templates for posters can be found online (see also Section 14.4).

### ➤ Make the title succinct, clear, and complete

The title should contain all important key terms and should grab the attention of your audience (see also Chapter 7, Section 7.5). It should be no longer than two lines. The font size needs to be big enough to be read about 10 feet away (1.5-in. minimum). The title is normally followed by the names of the authors and their affiliation.

### ➤ Make the Abstract no more than 50 to 100 words

Some posters include an Abstract. Others do not. If you are including an abstract on your poster, keep the abstract to a minimum. A good example of a possible Abstract follows.

**Example 14-1**

**ABSTRACT**

To examine the food requirements of cardinals, we have assessed the feeding frequency as well as the amount and type of food taken at feeders. We found that cardinals require on average 20 meals of 25 g a day during spring, summer, and fall. Their preferred food source consists of black sunflower seeds. During winter, birds needed about 33% more food per day. When feeding chicks, males and females both required 55% more seeds. Thus, it is important to offer more food to cardinals during winter and when they are rearing chicks.

Avoid using an Abstract on your poster that is a reprint of what you submitted to the conference committee, such as in the next example. No one in your audience wants to read over this entire, text-heavy Abstract, and the information contained therein can easily be shown under different and more visually appealing headings, such as the Introduction, Results, and Conclusions.

### ➤ Include a concise Introduction

Use the minimum of background information and definitions to get your viewers interested in the topic (no more than 200 words). Omit unnecessary details. During the poster session, you will be there to fill in details if needed. As most posters are displayed long after a meeting, the information on them must also be self- explanatory.

Ensure that your introduction contains all necessary elements (background, unknown/problem, question/purpose, experimental approach; see also Chapter 7, Section 7.7). These elements should also be signaled clearly.

**Example 14-2**

Background

Problem

Question/
purpose

Approach

**INTRODUCTION**

The landscape of the Middle East has been altered by human activity for most of the Holocene period. The rate of these modifications has accelerated in the last century, and today rapid population growth, political conflict, and water scarcity are common throughout the area. All of these factors increase the region's vulnerability to potentially negative impacts of climate change while decreasing the likelihood of successfully emerging region-wide adaptation strategies. In this study, we analyzed climate change in the Middle East during the 21st century as predicted by 18 Global Climate Models. The simulations were run as part of the Intergovernmental Panel on Climate Change Fourth Assessment Report (IPCC AR4) and used the Special Report on Emission Scenarios (SRES) A2 emission scenario, which is the scenario closest to a "business as usual" scenario in the SRES family.

*(With permission from Roland Geerken, modified)*

## ➤ Summarize your experimental approach only very briefly

For your Materials and Methods section, do not include as many details as you would in a research paper unless what you are presenting is a novel or unusual method. Aim for less than about 200 words. Consider using a flow chart or schematic to summarize experimental procedures if possible (Example 14-3b) because these are much more visually appealing and quicker to grasp than describing the same experimental procedure in words (Example 14-3a).

**Example 14-3a**

**UV mutagenesis of *Physcomitrella patens* to generate auxin mutants**

To isolate auxin regulated and signaling genes, our laboratory has generated UV- induced mutants by exposing 7-day-old *Physcomitrella patens* moss tissue, which had been blended for 2 min in a commercial blender after dilution of 1:100 (W/v), to 30 s of UV light. After UV light exposure, we prepared protoplasts as described in [2]. Protoplasts were replated and grown for 5–7 days under standard conditions on medium containing 0–2 μM NAA. Phenotypic mutants were selected visually based on abnormal branching or bud formation. Subsequently, mutants were screened as described in [3] and classified as reported previously [4].

**Example 14-3b**

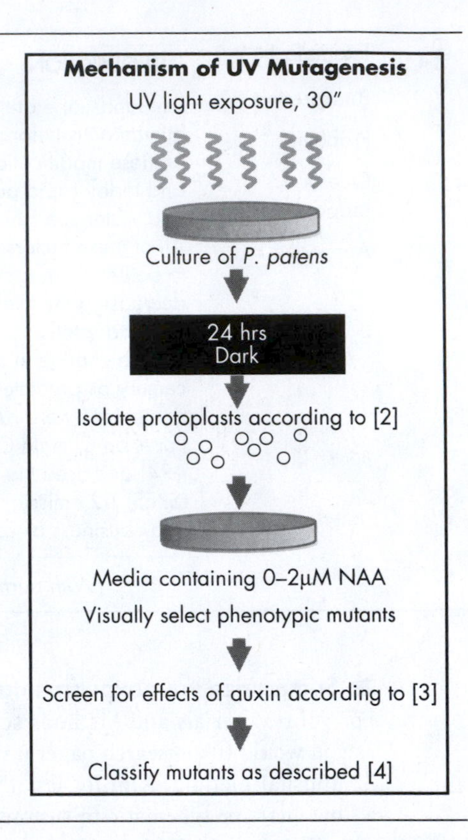


**➤ Present results mainly in figures and tables**

This section is usually the largest portion of the poster. Here you should describe your most important, overall results. Most, if not all, of your findings should be presented in the form of figures and tables. For ease of reading and comprehension for your viewers, ensure a consistent order between your results and the conclusions.

**Example 14-4**

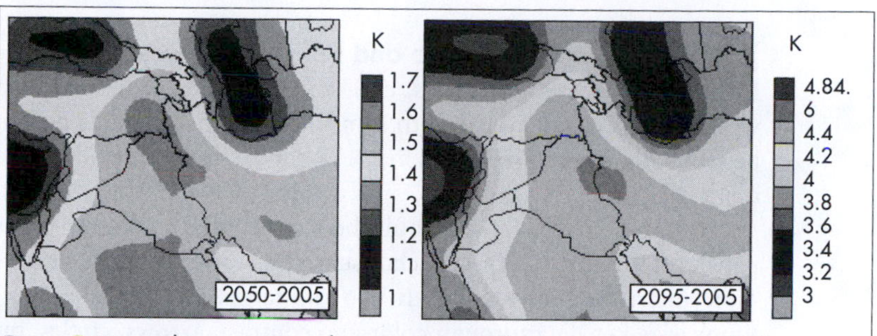


Figure 3. Mean change in annual temperature and precipitation. There is high agreement amongst the Global Climate Models for the predicted temperature change and significant disagreement for the predicted precipitation change.

*(With permission from Roland Geerken, modified)*

### ➤ In the Conclusions section, concentrate on your main findings and their meaning

Usually, conclusions are very brief because the poster is indicative of only a portion of the research. Mention only two to four main points in your conclusion or summary. If written as bullet points, these findings will be more visually pleasing than a whole paragraph of text.

**Example 14-5**

**CONCLUSIONS**

- Mean annual temperatures will increase by ~4 K by the late 21st century.
- Changes in precipitation are more variable; the largest change, however, is a precipitation decrease that occurs over an area covering the Eastern Mediterranean, Turkey, Syria, Northern Iraq, Northeastern Iran, and the Caucasus.
- Changes in precipitation will have a significant impact on fresh water resources.

*(With permission from Roland Geerken, modified)*

### ➤ Keep References and Acknowledgments to a minimum, and keep those included brief

References should be included if the technique is someone else's. Use as few references as possible. Usually, the use of names, dates, and journal information is enough for the Reference List. If applicable, thank individuals for *specific* contributions to the project, and mention any grant sources (i.e., summer research stipend, faculty grant, etc.).

### ➤ Make illustrations simple and self-explanatory

Include all the figures, graphs, and tables that you will want to point to when discussing your work with someone. This material in a poster should be self-explanatory. Thus, each figure and table should include a legend and a title.

Graphic material also should be simpler than material in published papers. If you have a choice, choose graphs rather than tables—they are more interesting for readers. If you use tables, make them very simple. Emphasize key data through color and contrast, arrows, circles, pull-outs, and other highlighting.

### ➤ Print out a miniature version of your poster

Once created, posters can be printed out from commercial printers. If your department or school does not have a poster printer, you can send your file to an online company that prints posters and ask them to mail it to you. Before printing as a poster, however, create a PDF file and print a miniature version of your poster on letter-sized paper to check layout, colors, font size, and design. You may also want to have miniature versions available as handouts during your poster session.

### ➤ Distinguish between poster and conference abstracts

If you are planning to present a poster at a conference, you typically have to submit an Abstract first. This Abstract will be reviewed by a committee, and only if accepted can you show your poster. If your Abstract is accepted, review the poster guidelines before starting to make the poster. Note that depending on your discipline, poster abstracts are usually not the same as conference abstracts. Poster abstracts tend to be much longer and more detailed as in the following example.

**Example 14-6**

**ABSTRACT**

**Isolation and Classification of Developmental *Physcomitrella patens* Mutants**

Mutants of *P. patens* are useful for the isolation of auxin regulated and signaling genes and can easily be generated by treating either spores or protonemal tissue with UV light. We have collected several UV-induced phenotypical mutants of the thiamine auxotroph ThiA1 of *P. patens* after UV exposure. These mutants were screened repeatedly for any auxin effects by growing them on media containing different auxin concentration (0–2 µM NAA). So far we have identified six different classes of mutants:

CLASS A: bud$^-$; these mutants do not produce buds in the first 3 weeks of growth under standard conditions, but are rescued by auxin in respect to bud and gametophore production (col T- 42, spa T- 33, spa T- 30, col T- 73, col T- 83)

CLASS B: bud$^-$; these mutants never produce buds or gametophores and are not rescued by auxin addition at concentrations of 0.025 to 2 μM for at least the first 3 weeks (spa T- 8, col T- 78)

CLASS C: gametophore growth not inhibited by auxin (up to 1 μM) (col T- 73)

CLASS D: filament length and number reduced (spa T- 3, col T- 25)

CLASS E: not inhibited by auxin in length and number of filaments (spa T- 8, spa T- 33, col T- 42, spa T- 30?, col T- 73?)

CLASS F: abnormal branching (spacing and rate) of caulonema; rescued by auxin (col T- 9, col T- 40)

The most interesting mutants so far appear to be the bud$^-$ mutants described in class A and B. Auxin in the cell has two possible origins: it is either made by the cell and then relocated, or transported into the cell via a putative auxin receptor after it was produced by a different cell. Auxin within the cell is normally present as conjugates with other molecules and is converted to the active IAA form when needed. Both Class A and B mutants were defect in the production of buds/gametophores. Since Class A mutants can be rescued by auxin, they cannot be blocked in the receptor or the conjugate or effect pathways, but must be blocked (at least partially) in auxin production and/or signaling genes. Class B mutants, on the other hand, are not rescued by auxin and can therefore be blocked at several positions: transport, conjugation, effect pathways, or signaling genes.

## 14.3 PRESENTING A POSTER

### ➤ Be prepared to tell viewers about your work and to answer questions

Although the material you are presenting should convey the essence of your message, make sure you are at your poster during your assigned presentation time to be available for discussion. In addition, ensure that you have prepared a 5- to 10-minute talk, highlighting the key points of your poster (see Chapter 13 for how to structure a talk). Practice this talk in front of your colleagues or professor. Your task as the presenter is also to answer questions as they arise.

During the actual presentation, focus on your graphics. Use your poster as a visual aid, but do not read it. Tell viewers the context of your research problem and why it is important, the objectives and how you achieved them, as well as the data and its significance.

## 14.4 USEFUL RESOURCES

For the top 20 tips in using PowerPoint, see Appendix D. For more suggestions and advice on how to create a poster, see also the following websites (last accessed February 2018):

- **http://www.swarthmore.edu/NatSci/cpurrin1/posteradvice.htm**—provides excellent advice on designing scientific posters written from a student point of view
- **http://www.ncsu.edu/project/posters**—gives detailed overview of planning, layout, editing and software
- **http://guides.nyu.edu/posters**—provides a good overview on creating a poster
- **http://www.makesigns.com/tutorials/**—lets you view video tutorials of how to create a scientific poster
- **http://posters4research.com/templates.php**—great site for free templates
- **http://www.utexas.edu/ugs/our/poster**—shows numerous sample posters and also contains templates

## 14.5 COMPLETE SAMPLE POSTER

Example 14-7 shows a well-designed poster with a short Introduction and Materials and Methods section, clearly-laid-out Results section with plenty of visuals, and a short, bulleted Conclusion.

**Example 14- 7**

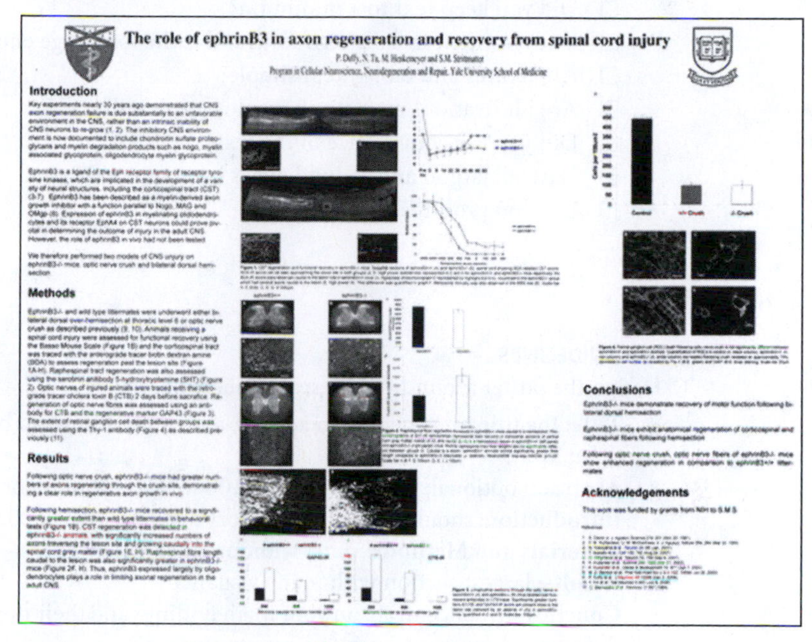


*(With permission from Philip Duffy)*

## 14.6 CHECKLIST FOR A POSTER

Use the following checklist to ensure that you have addressed all important elements for a poster:

- ☐ Do the illustrations tell the story?
- ☐ Is the purpose of the research or topic stated precisely?
- ☐ Is your poster abstract—if included—very short and different from your conference abstract?
- ☐ Does the Introduction have the following components?
  - ☐ Background
  - ☐ Problem or unknown
  - ☐ Purpose/topic or review
  - ☐ Overview of content
- ☐ Did you concentrate on the main points in each section?
- ☐ Is the flow of the panels self-evident to the viewers?
- ☐ Is the topic summarized and interpreted in the Conclusion section?
- ☐ Do all figures and tables have a title and a legend?
- ☐ Is your poster layout uncluttered?

- ☐ Did you use visuals where possible rather than text?
- ☐ Did you keep text to a minimum?
- ☐ Is text written in sans serif font, and is the font large enough?
- ☐ Are figures and tables kept simple?
- ☐ Are illustrations attractive? Is color used well?
- ☐ Did you use active voice in the text?
- ☐ Have all jargon and redundancies been omitted?
- ☐ Did you proofread your text?

## SUMMARY

### POSTER GUIDELINES

1. Design the poster around your research question. Include:
    - **Title:** the title must be succinct, clear, and large enough to be read 6 feet away.
    - **Abstract:** optional; fewer than 50 to 100 words if you use one.
    - **Introduction:** should be self-explanatory and fewer than 200 words.
    - **Materials and Methods:** short without much details.
    - **Results:** largest section with many visuals.
    - **Conclusion:** concentrate on your main findings and their meaning.
    - **References:** optional; keep to a minimum.
    - **Acknowledgments:** optional; keep short.
2. Find visual ways to show your work—let the illustrations tell the story.
3. Focus only on the main points in each section.
4. Aim for about 20% text, 40% graphics, and 40% empty space.
5. Lay out the poster in a logical, easy-to-follow way.
6. Design backgrounds well.
7. Make fonts large enough to read.
8. Make illustrations simple and self-explanatory.
9. Print out a miniature version of your poster.
10. Distinguish between poster abstracts and conference abstracts.
11. Be prepared to tell viewers about your work and to answer questions.

## PROBLEMS

### PROBLEM 14-1 Poster

**Create an attractive poster from one of the following:**

1. A set of related laboratory experiments.
2. A paper you have selected from an academic journal.

**PROBLEM 14-2 Poster**

**Attend a poster session, and choose a poster for review. Ask the presenter questions about the work, and then write a brief critique of the poster.**

**PROBLEM 14-3 Poster**

**Why is the following not a good Abstract for a poster? Rewrite the Abstract so that it could be included on a poster.**

Certain plants have adapted to desert environments and actually thrive in it. Some desert plants are able to absorb large quantities of water in a short time and store it in roots, stems, or leaves. Others are drought-tolerant or drought-evasive. Yet others combine these characteristics. Drought tolerance is of great interests to scientists because drought-tolerant plants can withstand long periods without water when they become dormant. However, it is not well understood under what conditions dormancy ends and is induced. We studied a drought-tolerant plant, brittlebush (*Encelia farinosa*), to understand when and how these bushes grow new roots and leaves and reestablish full metabolic activity. Dormant plants were studied on site by adding various amounts of water to their root system. We noted depths of the water soaking into the soil, the time it took for roots and leaves to sprout, and the length of time before leaves were dropped again when water was withheld. We found that a minimum of a half inch of rain is required for brittlebush to come out of dormancy. Our results suggest that water needs to reach the deeper roots of the plants before dormancy is broken, and the longer the moisture is retained in the soil, the longer growth of the bushes is sustained.

**PROBLEM 14-4 Poster**

**Would the following be an engaging panel on a poster? Why or why not?**

**Conclusion**

Here we report that recombinant bindin retains the ability to agglutinate eggs species-specifically. We have produced a series of amino- and carboxyl-terminal deletion analogs of *S. purpuratus* bindin to investigate the minimum structure necessary for agglutination. We examined the species specificity of the resulting mutant proteins in egg agglutination assays. Results of these investigations indicate that either the amino- or carboxyl-terminal regions flanking the conserved central domain is sufficient for species specificity. Both ends contain repeated sequence elements that are different in bindins from different species.

CHAPTER 15

# Research Proposals

THIS CHAPTER DISCUSSES:

- The components and format of grant proposals
- Individual sections of a grant proposal
- The importance of the first sentence and the overall objective

## 15.1 GENERAL OVERVIEW

To be successful, scientists require funds and have to find sources for this money. These funds are usually garnered through successful grant applications. Grant writing is therefore as central to a scientific career as is writing papers and doing research at the bench.

Before composing a proposal, you must research a funder's priorities. Your proposal must establish a connection between your project's goals and the agency's interests. The proposal must succinctly but thoroughly present the need or problem, the proposed solution, and your organization's qualifications for implementing that solution. It must be drafted with much care to boost your chances of success.

## 15.2 COMPONENTS AND FORMAT

### ➤ Follow instructions EXACTLY

If you have instructions for your research proposal, follow them EXACTLY. Sometimes, proposals receive a bad evaluation grade simply because they do not conform to the required format specified by the instructions. There are many funders, from federal to private, and most have their own ideas on how they like to have proposals structured. There are even more people seeking funding. Your competition will likely be tough. To maximize your chances of obtaining funding, it is therefore important to present information as requested.

Generally, your proposal should answer the following questions:

- Why this project?
- Why you?
- Why at your institution?
- Why this sponsor?
- Why now?

**➤ Include the following sections in a proposal:**

- **Abstract**
- **Specific aims**
- **Background, including Statement of Need**
- **Research Design**
- **Significance/Impact**

These core sections are contained in every proposal, but may be called differently. Some proposals may include additional sections, such as a Budget or Innovation section. In some proposals, the Significance/Impact section appears before the research design section.

## 15.3 ABSTRACT

**➤ Write a concise abstract that includes all the main points of your proposal:**

- Background
- Unknown or problem/statement of need
- Overall objective of the proposal
- General strategy
- Significance/impact

The Abstract summarizes the proposal, articulating the highlights from each section. It includes all of the main information covered in the proposal (background, unknown or problem/statement of need, overall objective, general strategy, and significance/impact) in a single paragraph. The Abstract must be written such that it can stand on its own, without the detailed narrative. It must be concise, informative, and complete.

**➤ Pay attention to the first and last sentence**

Pay special attention to power positions in your Abstract. Put much effort into writing the first sentence. Start it with an important key term, and keep the sentence short. It provides the first impression to the reader and should introduce the topic of the proposal. Almost as important as the first sentence is the last one. It should highlight why the proposal is important to society, and thus contains the Significance or Impact statement. The abstract in Example 15-1 starts with a strong first sentence and ends with a great closing sentence. It is logically constructed and presents individual elements in the order most reviewers expect to find them.

---

**Example 15-1** **Proposal abstract with integrated specific aims**

Background
Problem
Objective and specific aims
Strategy
Significance

Plate tectonics distinguishes Earth from other terrestrial planets. Plate tectonics arise due to the convection in the mantle of the Earth, and for plate tectonics to occur, some mechanism must exist to compensate temperature-dependent viscosity. However, it is not well understood what this mechanism is. In this proposal, we aim **to approach this long-standing mystery through employing a MCMC algorithm that will (1) systematically explore various sampling strategies and (2) seek the fastest forward model calculation by benchmarking competing Stokes flow solvers, including the pseudo-compressibility method.** To lay a foundation for realistic 3-D applications, all computations will be conducted in the 2-D formulation. Microsoft Windows OS will be the main platform for developing the Monte Carlo code and the flow solver as well as for analyzing and visualizing the results of MCMC simulations. Stokes flow calculations will be done with the PI's ABC server. Understanding the physics of plate-tectonic convection in the Earth's mantle will have profound implications for our understanding of the habitability of a terrestrial planet and the evolution of life.

*(Jun Korenaga, proposal to private foundation; modified)*

---

## ➤ Include a statement of need

The statement of need is a key component of any proposal. If it is not included, the proposal will not be convincing to readers. Readers will not find any important reason(s) for the proposed project as the need or problem that has to be met is not laid out clearly.

---

**Example 15-2** **Statement of need**

Salamander limb deformities detected in California are of concern to public health experts. It is important to explore the underlying causes of these deformities to establish preventive measures. An urgently needed next step in the protection of public health is the examination of the X hypothesis. Investigations of this hypothesis may not only determine the causative agent of the observed limb deformities but also forgo the danger of potential developmental problems in humans.

---

## ➤ Clearly identify the overall objective

Within the Abstract, clearly identify the overall objective of your proposal so that the readers understand your goal for the proposed work. The objective is the most important element of your proposal. It should follow logically from the statement of need or the problem. Ensure that reviewers can immediately identify this statement. Either place it in italics or boldface or label it, for example, "Objective" or "Goal."

The overall objective can be subdivided into specific (short-term) aims, which are the two to five steps needed to achieve your objective in the time frame of the proposal (usually 2–5 years). Specific aims can be listed within the abstract following the objective or as a separate section.

## 15.4 SPECIFIC AIMS

### ➤ List your specific aims in precise language

The Specific Aims section may be a separate section by itself or may be combined with the Abstract. Reviewers will read this section very carefully, as it serves as an orientation for the rest of the proposal and provides the detailed approach to your overall goal laid out in two to five concrete steps.

Your specific aims should be written in a list rather than in paragraph form, and they should be placed in boldface to highlight them. Each specific aim may be followed by a brief (one paragraph) description of the proposed approach. This description typically summarizes the preliminary results or rationale, the hypothesis, and the approach for each aim. If a hypothesis statement is included, consider placing it in italics to set it apart. As a final sentence in this narrative, you may also state expected outcomes for the specific aim.

**Example 15-3** **Specific aims**

*Overall objective: to identify and purify natural chemicals of oceanic plants and to test their activity as potential medicines.*

**Specific Aim 1. Identify plants in Florida Keys that contain compounds with antimicrobial properties.** In preliminary work, we have found that extracts taken from three plants inhibit growth of bacterial cultures. We will now screen and assess additional plants in the Florida Keys for potential antimicrobial agents by using microbial techniques and chemical fingerprinting. This aim will allow us to collect a number of extracts for further analysis in Aims 2 and 3.

**Specific Aim 2. Isolate and characterize antimicrobial compounds from plants identified in Aim 1 by . . .,**

## 15.5 BACKGROUND

### ➤ Provide context and preliminary results for each aim

Follow the Abstract/Overview paragraph with background information, which may also include important preliminary results. The Background section provides context for the readers by describing the current state-of-the-art, or what is presently known about the topic. It can be written as one section or split into subsections (largely based on aims). This section has to convince reviewers that specific aims, once achieved, will have significant impact on the topic in question. Do not just state that you aim to gain scientific knowledge in your field but rather tie the proposal in to some broader scientific or clinical picture.

Generally, readers expect the parts of the Background section to be arranged in a standard structure:

**Background:**
  **Within subsections:**
    **Background/known**
    **Unknown/problem/need**
    Aim/hypothesis (optional)
  Summary (optional)

The following example starts with general background to provide context and then lists the problem and offers the hint of a possible solution by generally reporting on some preliminary results:

**Example 15-4 Background funneling to unknown**

| | |
|---|---|
| Context/ background | Global warming is arguably one of the most pressing concerns of our time. It has been linked to the rapidity of observed climate change—the fact that the Earth's temperature rose by approximately 0.7°C over the last century (the most dramatic increase documented in historic times) and the attendant threat posed by melting polar icecaps, rising sea levels, and potentially, more severe weather patterns. We do not know yet what proportion of this global warming is due to human activity and what is due |
| Specific problem/need | to natural variations. More important, we lack an effective model to predict precisely by how much the temperature will rise as a consequence of the increase of the levels of $CO_2$ and other greenhouse gases in the atmosphere of the |
| Specific aim | Earth. In this proposal we aim to . . . |

## 15.6 RESEARCH DESIGN

### ➤ Describe the approach for each specific aim in detail

The Research Design section is also sometimes referred to as the Experimental Approach, Methodology, or Strategy. The main function of this section is to propose in detail what approach you will take to address each of your specific aims over the course of the proposed project. This section, which typically is also the longest section of a proposal, will receive the most scrutiny during review because it is the heart of your proposal. It should lay out a sound and attainable approach to address the stated problem or need.

To delineate your plans convincingly, the section should elucidate how you have arrived at your findings, what you expect to add to the topic, the specific experimental approaches, the expected outcomes of your experiments, potential alternate approaches, and the implications your project will have.

### ➤ Organize the section into subsections according to your specific aims

To help reviewers find the corresponding approach for each aim, divide the Research Approach section into subsections based on your specific aims. Give these subsections the same title as those for your specific aims. Within the subsections, cover:

**Heading: Specific aim/objective**
Rationale/hypothesis (not always required)
**Experimental design**
**Analysis**
**Expected results (outcomes and significance)**
Alternative strategies (not always required)

---

**Example 15-5** **Research Design and Methods section**

**I. Laboratory and Field Behavior Study**

Objective: To find natural insect predators of the Monarch butterfly, *Danaus plexippus* L., in the California Central Valley and the Sierras.

Approach and Analysis: For our experiment, we will tether live Monarch butterflies to diverse vegetation in various locations throughout the Central Valley and the Sierras to observe what insect predators prey on Monarch butterflies. Vegetation will include various chaparral plants, cacti, as well as trees, including oak, pine, fir, and palm trees. Butterflies will remain tethered to vegetation for 24 hours. Different heights of tethering will be recorded, and the exact location and timing will be noted. A total of 1,000 butterflies will be used over a 1-month period in the spring at both locations. Using high speed video camera set on the butterflies, any attack on the butterflies will be recorded over the 24-hour period. Data will subsequently be analyzed as to the time before a butterfly was attacked and the type of predator.

Expected Outcome: We expect to obtain information as to what insect predators attack Monarch butterflies and how quickly predators find and attack the butterflies. Findings will give valuable insights into predation dangers butterflies encounter during part of their migration journey through California and provide baseline data to which future findings can be compared.

## 15.7 IMPACT AND SIGNIFICANCE

### ➤ End the proposal with a broad impact statement

In the last paragraph of your proposal, state the impact and significance of the overall project. Focus on how human beings will benefit rather than on your own goals. You may also mention evaluation plans such as measurable objectives. The following example shows a typical impact statement.

**Example 15-6** **Impact and significance**

Novel, genetically modified crops will enable farmers to grow various assortments of plants with increased productivity, quality, and enhanced adaptation and survival capabilities under adverse environmental conditions. The need for such crops is especially critical given the global food crisis, which puts nearly a billion people at risk of hunger and malnutrition. Using new technologies and cultivating varieties that permit green farming, the proposed program intends to provide novel and environmentally friendly answers to the challenges crop growers experience globally.

## 15.8 ADDITIONAL PROPOSAL COMPONENTS

Depending on where you send your proposal, you may also be asked to include additional information other than the ones discussed so far. Such information may range from CVs and transcripts to reference letters, certifications regarding conflicts of interest, conduct of ethical research, animal experimentation, guarantees of permission to conduct research in other countries, and so forth. Another very important section found in most proposals is the budget, a detailed breakdown of the financial support requested from the sponsoring agency.

### ➤ Costs should be realistic and justified

Conducting research costs money, often more than one imagines. Expenses arise, for example, for materials, equipment, personnel, travel, and facilities.

Usually, you have to list the expenses required to successfully perform the proposed research. Your business office typically assists you in putting such a budget together to ensure that costs are realistic and justified. A sample budget is shown in the following example:

**Example 15-7** **Short sample budget table**

**Use of Funds**

The total project cost is $xx,xxx. The requested funds are allocated as follows:

| | | |
|---|---|---|
| Personnel | Principal investigator, salary for 2 months | $xx,xxx including benefits |
| | Postdoctoral fellow | $xx,xxx including benefits |
| Equipment | Laptop computer | $x,xxx |
| Travel | Domestic and foreign | $xx,xxx |
| Publication costs | | $x,xxx |

## 15.9 CHECKLIST

When you have finished writing the proposal (or if you are asked to edit these sections for a colleague), you can use the following checklist to "dissect" the sections systematically:

- ☐ Is the proposed topic original?
- ☐ Are all the components there? To ensure that all necessary components are present, in the margins of the proposal clearly mark the following:
  - ☐ Abstract/overview
  - ☐ Introduction/background
  - ☐ Research design
  - ☐ Impact/significance
- ☐ Is the Abstract kept short and within the set limits?
- ☐ Does the Abstract contain:
  - ☐ Context/background (optional)
  - ☐ Statement of need
  - ☐ Objective
  - ☐ Strategy
  - ☐ Significance/impact of initiative
- ☐ Is the overall objective stated precisely?
- ☐ Is the unknown/problem clear?
- ☐ Is the approach clearly delineated, and does it include a rationale as well as expected outcomes and significance?

## SUMMARY

### PROPOSAL GUIDELINES

1. Follow instructions EXACTLY.
2. Include the following sections in a proposal:
   - **Abstract:**
     - Include: background, unknown/problem/need, **overall objective**, general strategy, and significance/impact.
     - Pay attention to the first and last sentence.
     - Include a statement of need.
     - Clearly identify the overall objective.
   - **Specific Aims:**
     - List your specific aims in precise language.
   - **Introduction/Background:**
     - Include: background, statement of need, aim/hypothesis (optional).
     - Provide context and preliminary results for each aim.
   - **Research Design:**
     - Include: rationale/hypothesis (not always required), experimental design analysis, expected results, alternative strategies (not always required).
     - Describe the approach for each specific aim in detail.
     - Organize the section into subsections according to the specific aims.
   - **Significance/Impact:**
     - End the proposal with a broad impact statement.
3. Costs should be realistic and justified.

## PROBLEMS

### PROBLEM 15-1

**Write a short proposal on a project of your interest or one given by your instructor. Work in groups of three to five to evaluate and score each other's proposals. Evaluate proposals base on clarity of writing, placement of information, and how convincing and realistic a project is portrayed. Also consider whether the budget, if included, is realistic.**

### PROBLEM 15-2

**What is the problem with the following Abstract submitted for a research proposal? Rewrite the Abstract such that it would be acceptable for a proposal.**

**Background:** Cardiovirus is a common cause of gastroenteritis. For routine vaccination of Chinese infants, a new cardiovirus vaccine has been recommended.

**Objective:** To evaluate the impact and cost-effectiveness of the Chinese cardiovirus vaccine program using a dynamic model of cardiovirus transmission.
**Expected outcome:** Our analysis will indicate the impact and effectiveness of a rotavirus vaccination program.
**Significance:** Findings can be used to inform policy makers to prevent rotavirus infection.

**PROBLEM 15-3**

**Identify all components of the following proposal abstract/overview (background, need/problem, objective, specific aims, approach, impact):**

**Abstract**

The role that soils play in mediating global biogeochemical processes is a significant area of uncertainty in ecosystem ecology. One of the main reasons for this uncertainty is that we have a limited understanding of belowground microbial community structure and how this structure is linked to soil processes. Building upon established theory in soil microbial ecology and ecosystem ecology, we predict that the structure of belowground microbial communities will be a key driver of carbon and nutrient dynamics in terrestrial ecosystems. We propose to test and develop the established theories by combining state-of-the-art DNA-based techniques for microbial community analysis together with stable isotope tracer techniques. By doing so, we expect to advance our conceptual and practical understanding of the fundamental linkages between soil microbial community structure and ecosystem-level carbon and nutrient dynamics.

*(Mark Bradford and Noah Frierer, proposal to private foundation)*

**PROBLEM 15-4**

**Point out the problems in the following impact statement.**

**Impact/Significance**

Funding from the Foundation will not only provide for my postdoctoral fellow but also will result in the publication of two papers over the funding period.

**PROBLEM 15-5**

**Evaluate the following short research proposal. Is it clear what the authors are suggesting to do?**

Sea urchin gametes serve as a model system in fertilization studies. Their gametes aid in answering important questions in the field and, subsequently, in transferring this knowledge to fertilization interactions of even higher organisms such as mammals.

There are several advantages to using urchin gametes. The biological significance of urchin gamete interaction is well established for cell-cell adhesion. Gamete interaction occurs rapidly and synchronously. Gametes are homogeneous populations of single cells and are readily available in large quantities.

The natural habitat of the sea urchin species *Strongylocentrotus purpuratus* and *S. franciscanus* is along the Pacific coastline from Canada to Baja California. *S. purpuratus* are found in the intertidal zone, while *S. franciscanus* normally prefer somewhat deeper water. However, the two species are often intermingled just below the low water mark. They both have overlapping breeding seasons, but fertilization appears to be largely species-specific. Spawning occurs in December through March, depending on the location.

The urchins can be collected when they are ripe to obtain their gametes. When the urchins are injected with KCl, they spawn and eggs can be collected for fertilization and egg agglutination experiments; sperm can be collected for fertilization and sperm adhesion studies as well as for the isolation of the protein bindin, thought to play a critical role in fertilization and species specificity.

During fertilization, species specificity is determined by the following events: The sea urchin egg is coated by a thick jelly layer, which consists largely of sulfated polysaccharides. The egg plasma membrane is also surrounded by a thin layer called the vitellin layer. The sperm have to pass through the jelly layer in order to adhere to the vitellin layer and fertilize the egg. When the sperm come into contact with the jelly layer, the acrosome reaction is triggered. During the acrosome reaction, the membrane of the acrosomal granule fuses with the overlaying sperm plasma membrane. Bindin is exposed after exocytosis of the granule. Bindin adheres to a vitellin envelope receptor, which contains sulfated, fucose-containing polysaccharides. Extension of the acrosomal process occurs, with bindin coating it on the outside. Then, the acrosomal process penetrates the vitellin envelope. Finally, fusion of the plasma membrane of the sperm and egg occurs.

We are particularly interested in the protein bindin and its structure-function relationship. Specifically, we are investigating how bindin adds to the specificity of sea urchin fertilization. Key questions we are interested in answering include: Is bindin the only sperm protein involved in determining species specificity? What is bindin's receptor on the egg surface? What specific portion of the bindin protein is responsible for species-specificity?

Aside from its role in sperm adhesion, bindin also seems to aggregate in larger clusters. These clusters can cause sea urchin eggs to stick together, and this interaction is species-specific. It may be possible to use this fact as an assay for determining activity of the protein.

In summary, sea urchin fertilization allows us to gain important insights into gamete interactions and specificity. Learning more about the protein bindin and its role in this process will poise us to search for similar proteins and interactions in mammalian cells.

CHAPTER 16

# Job Applications

THIS CHAPTER DISCUSSES:

- Curricula vitae and résumés
- Application-accompanying documents: cover letters and personal statements
- Candidate selection
- Letters of recommendation

## 16.1 CURRICULA VITAE AND RÉSUMÉS

To enter into a successful career, you need to be familiar with the appropriate documents necessary to take this step. Your curriculum vitae, commonly referred to as a CV, or your résumé is one of the first items that a potential employer sees. Both of these documents should be updated and maintained on an ongoing basis. Know that good writing requires plenty of time to complete. You will not be writing just one draft of your CV/ résumé (and other application material). You will need to work on these documents over time. So, start early!

### ➤ Curricula Vitae

A CV is a summary of your educational and academic backgrounds and is used primarily in academic and medical settings. CVs are usually very comprehensive and elaborate on professional history, including every term of employment, academic credential, publication, contribution, or significant academic achievement. Do not underestimate your skills or overlook those you may have learned and practiced in lab courses. Typically, CVs have no length restrictions, but it is essential that your CV is clear, concise, and honest.

### Tailor your CV specifically to the position and to the organization

Customize your CV for each job application as well as for each grant or fellowship application. That is, organize and present your information based on the interests of the employer, highlighting your main qualities on the first page (or two). It has to be apparent that you are the person best matched to the position in terms of skills, training, experience, and interest. If your CV is tailored to the position and company, you stand a much better chance of getting an interview and being hired. If needed, do your homework and obtain pertinent information on this employer and the position to which you plan to apply. Note that some institutions and organizations have established formats for CVs. Before applying for a position, be sure to check their websites for templates and instructions.

### Highlight your academic achievements in your CV

Because CVs are geared primarily toward the medical sector and academia, it is important to emphasize your academic achievements. Thus, include publications, talks, teaching and research experiences, honors and awards, and memberships in honor societies. Realize, however, that CVs differ widely, depending on where you are applying, your experience, and your personal preference. The following list gives you an example of what categories you may include in your CV:

- Name and address (no personal information, such as birth date, social media address, or social security number)
- Education, with degrees and dates (most recent first)
- Clinical certifications, with dates
- Employment history, with brief description and dates (most recent first)
- Grant funding
- Leadership and service
- Laboratory skills
- Publications—name in bold, co-first authors easily identified
- Invited presentations and seminars
- Teaching experience
- Professional qualifications
- Certifications and accreditations
- Computer skills
- Language proficiency
- Unique technical abilities
- Professional memberships
- Honors and awards

Following is a sample CV that serves as an example of the appropriate format for such a document. Note that educational qualifications and work experience are listed in reverse chronological order.

 **Example 16-1** **Sample CV**

Name and contact information can easily be found at the top.

This CV is free of misspellings and grammatical errors.

The applicant includes information on education followed by other academic achievements in order of importance.

The formatting categorizes and highlights information.

Dates, lists, and subsections are perfectly aligned, making the CV look neat.

Items are listed in reverse chronological order for ease of finding the most important one(s) at the top.

The CV ends with references, which are made available on request.

**JANE SMITH**

83 River Road, Huntington Beach, CA 92615

(714) 557 9003

Jmill3@yahoo.com

**EDUCATION**

2014—present *Washington University*, St. Louis, MO; Major: Biology; BS expected June 2015; current GPA: 3.15

2010—2014 *Hand High School*, Madison, CT; GPA: 3.2

**SENIOR PROJECT**

Chemical composition of maple syrup produced by three different species of maple in New England

**TALKS**

*Maple Syrup Compositions.* VIIth International Undergraduate Research Conference. Woods Hole, Massachusetts 2017

**RELEVANT COURSEWORK**

Molecular Biology

Biochemistry I and II

Organic Chemistry I and II

Inorganic Chemistry

Undergraduate Research Volunteer

**MEMBERSHIPS**

***American Society for Cell Biology (ASCB)*** Since 2017

**AWARDS AND HONORS**

***Washington University Provost's Award for Excellence in Research*** Fall 2017

***John Hawkins Award*** for Undergraduate Research May 2017

***Phi Beta Kappa*** Spring 2016

**SPECIAL SKILLS**

Operation of HPLC and NMR

Solid computer skills (Word, Excel, PowerPoint, Adobe systems)

**REFERENCES**

Available on request

Note that you should always ask someone before you provide their name and contact information as a reference.

## ➤ Résumés

In contrast to a CV, a résumé contains only experience directly relevant to a particular position. Résumés are primarily used for seeking employment in the private sector. They include a summary or listing of your relevant job experiences and education.

### Tailor your résumé to the position and to the organization

Your résumé should be tailor-made for the position to which you intend to apply. It should be job-oriented and concise. Indeed, every new job application requires a résumé written for that particular job.

Typically, résumés for industry are usually shorter than CVs sent to academia, about one to two pages, depending on your job experiences. Unlike a CV, a résumé puts less emphasis on academic achievement and more on professional or work experience. Thus, listing topics such as technical skills will be more important than listing talks, abstracts, and memberships in honor societies.

### Highlight your skills and professional experiences in your résumé

Skills you may have acquired should be highlighted, for example, in a separate "Technical Expertise and Skills" section (see the sample résumé following for ideas and wording on what to include in such a section). You may be surprised what you can include in such a section based on what you may have learned during your time in college or university, particularly if you did any undergraduate research. However, be careful not to overstate your skills unless warranted. Aside from skills, highlight work experiences. Résumés may also contain broader experiences outside of work, such as military service or specific programs and courses you may have taken (AutoCAD, continuing education programs), if they are relevant to the position you are seeking. Look for ways to expand your education and training to set yourself apart from your peers—you will become more employable. You will also be more attractive to potential employers if you can show consistent and sustained engagement in their particular field of work.

Common mistakes in composing a résumé include:

- Not putting your best or most important and relevant experiences near the top
- Listing items in chronological instead of reverse chronological order
- Mentioning specific courses simply because you particularly enjoyed them rather than for their relevance to the job

Many employers request a one-page résumé. Although tight, this one-page limit helps streamline the content to only relevant information. You will need to be very selective on what to include and where in the document you place this information. Starting a second page to include a job as a camp counselor and a team sport is not worth the extra page unless you were on the college team, won a major award, or the like.

The next two examples show two sample résumés. The first one is from a student fresh out of college applying for her first job, the second one from someone who has held a job in industry already and is now looking for another opportunity.

**Example 16-2** **Sample résumé 1**

Name and contact information is provided at the top.

**MITCH A. MENDOL**

21 Washington Drive, Madison, CT 06419

1-860-544-4097

mmendol@gmail.com

The résumé shows that the applicant is a recent college graduate with a good grade point average (GPA). It also stresses his engagement and interest in field biology and environmental ecology.

***College graduate with research and lab management experience and excellent interpersonal skills seeking employment as a research lab manager. Outstanding attention to detail, work ethic, and service oriented.***

**EDUCATION**

**Bachelor of Science in Biology** **May 2016**

*University of New Hampshire, Durham, NH*

Minor in Anthropology

**GPA:** 3.3

*Highlighted Coursework: Undergraduate Research in Field Biology, Environmental Ecology*

The information section most relevant to the job posting immediately follows that on Education because this will be of great interest to the potential employer.

**TECHNICAL SKILLS**

Experience with environmental biology field work and data analysis (SPSS)

Skilled in media preparation and autoclaving, and preparing solutions and buffers

Skilled in Microsoft Office Suite, including Word, Excel, PowerPoint, and Outlook

Skilled in Windows and Mac operating systems

Proficient in basic Spanish

Important and relevant work experiences are listed. Here, too, it is apparent that the applicant had an interest in working in a research lab.

**EXPERIENCE**

**University of New Hampshire, Durham, NH**

*Lab Assistant* *October 2015–August 2016*

- Prepared media, buffers, and solutions for environmental lab
- Coordinated orders and maintained lab inventory; managed autoclave area
- Assisted faculty with student lab preparations

It is also apparent that he is a hard worker based on the summer jobs and semester jobs he held.

**The Works Bakery Café, Durham, NH**

*Preparation Cook, Baker, and Counter Associate* *Summer 2015*

- Coordinated food preparation, baking, and ordering

**Yale University School of Medicine Emergency Department, New Haven, CT**

*Research Assistant* *Summer 2014*

- Screened and interviewed patients; collected data on health and media use in coordination with researchers and hospital staff
- Managed online databases; uploaded and analyzed survey data

**University of New Hampshire Athletic Department, Durham, NH**

*Equipment Room Assistant* *Fall Semester 2013*

- Assisted with support of numerous varsity teams, including caring for uniforms and equipment
- Maintained and organized equipment inventory lists and packing checklists for away games

**Malone's Restaurant, Madison, Connecticut**

*Preparation Cook* *Summer 2012*

- Assisted in completion of menus and coordination of kitchen maintenance and cleanliness
- Processed and maintained orders of food

---

**Example 16-3** **Sample résumé 2**

The résumé provides the name and contact information at the top, where it can be found easily.

The résumé is free of misspellings and grammatical errors.

**RÉSUMÉ**

**Jorge Garcia Colon**

**CONTACT ADDRESS** Address; Telephone; email

**PROFILE**

- Highly motivated biochemistry mayor with experience in peptide synthesis and HPLC
- Excellent team player with good verbal and written communication skills

The most important and relevant research experiences and skills are placed at the beginning; the order of the remaining sections is by importance and relevance.

**ACADEMICS**

**University of Connecticut**, GPA: 3.75 2014–present

Major: Biochemistry

(Selected coursework to date: Junior-level biology courses, organismal behavior, statistics, organic and inorganic chemistry)

**Haddam-Killingworth High School**, Haddam, CT; GPA: 3.85 2010–2014

The applicant uses formatting to categorize and highlight information. He also lists all his skills clearly. In addition, he uses important key words and wording taken from the job advertisement to allow for machine searching.

**RESEARCH EXPERIENCE**

**Peptide synthesis research** Summer 2013

Yale University, Professor X's Lab

**SKILLS**

| | |
|---|---|
| *Biochemistry:* | Experience in peptide synthesis and HPLC |
| *Computers:* | Experience in Windows and Mac operating systems, Adobe Creative Suite 5, LaTeX |

The applicant lists his experience in reverse chronological order, putting the most important/relevant positions first.

He ensures perfect alignment and spacing of dates, lists, and subsections. This makes the résumé look neat.

**INTERNSHIPS AND WORK EXPERIENCES**

**Seho Systems GmbH, Germany** June—July 2016

Internship; accounting, payroll, record keeping, customer service, sales, and electronics pre-assembly

**Brauerei Martinsbräu, Germany** June—July 2015

Internship; assisted in brewing, accounting, and records keeping

**Technical Assistant, ABC University** 2015—2017

Assisted in lab organization and preparation; purchase orders of reagents

**SCHOLARSHIPS, HONORS, AND AWARDS**

UConn Leadership Scholarship 2014–present
Killingworth Ambulance Association Scholarship 2014
National Honor Society 2013–2014

**LANGUAGES**

**German** native speaker
**Spanish** native speaker; excellent oral and written Spanish language skills

---

### ➤ Ensure that your CV or résumé is well presented and flawless

People who review your application will carefully look at your CV or résumé as well as at the accompanying cover letter. They will not only evaluate your experience and skill level but may also note the layout, spacing, grammar, style consistency, and spelling of your text. Therefore, it is essential that the document is well prepared and free of errors.

### ➤ Format your CV or résumé strategically

The use of boldface, italics, spacing (margins and line spacing), and font can make a big difference in ease of reading and in finding information for anyone on a search committee. Therefore, do not put everything in the same type. Rather, format the document so that categories are clearly distinguishable and your CV or résumé stands out from others.

If you have to upload your CV or résumé in a text-only format on a website, you need to create a plain text format document. You will not be able to use boldface, italics, or bullet points, but if you do it right, you can get the document to look very similar to the formatted version. Save your existing CV or résumé in Microsoft Word (or similar word-processing program) as a plain text or Notepad format, then open the plain text file and clean it up, adjusting spacing and using * or > symbols for bullets, for example.

Be sure to include important key words that appear in the job posting. Sometimes CVs or résumés are read by machines that look specifically for these terms. This is another reason why your CV or résumé should be redone for every job application. To that end, be sure that you also have cleaned up or locked down your social media sites. More often than not, potential employers will look at these to garner more information about you. Realize also that specific states and countries have rules regarding security clearances, felony convictions, criminal records searches, and the like.

### ➤ Know where to get help

If you need help in composing any of these documents or would like to practice interviewing, go to a career service office. This type of office can be found on almost all campuses, and most will be happy to assist. Career service offices typically have résumé services and may even have business-appropriate interview clothes for students to borrow. Many instructors also can assist with advice. I encourage you to create and keep an electronic portfolio (e-portfolio) for your CV, training certificates, labs, capstone projects, conference programs, relevant faculty names for work studies (along with dates, names, and places), and so forth.

## 16.2 COVER LETTERS

### ➤ Customize the cover letter

When you are applying for a job, you will have to send not only a CV or résumé but also a cover letter. Your cover letter creates a professional impression from the outset. It allows the search committee to see your written communication skills. Like the CV or résumé, it should be flawless because it can make or break your application. It should also be pleasing to the eye and thus well placed on the paper.

In your cover letter, address concrete and specific requirements raised by the job posting where possible. Whenever possible, address your cover letter to a specific person, even if that means you have to call the organization to inquire about the name. Check (and double-check) the letter for spelling and grammatical errors. If you are sending a hard copy, print it onto good-quality paper. Preferably, do not send more than a one-page cover letter.

Following is an outline for the general organization of a cover letter:

Opening paragraph:
- State the purpose of your letter.
- Mention how you heard about the job.

Middle paragraph(s):
- Highlight your past accomplishments.
- Describe your goals.
- Explain why you believe you are a good fit for the position.

Closing paragraph:
- Mention any enclosures (CV, publication samples, personal statement, etc.).
- Make positive closing remarks.

Example 16-4 shows a sample cover letter that follows the outline given previously:

---

**Example 16-4** **Cover letter 1**

The letterhead provides the name and contact information of the sender.

**John Smith**

*345 Main Street*
*Plainfield, MA 01070*
*Email* *Telephone number*

To
Name
Address

Date

| | |
|---|---|
| The applicant uses the name and correct title for the person he is addressing. | Dear Professor Ying:<br>I am writing to apply for the **Technical Assistant position** as posted on University XXX's website. I have always been very enthusiastic about working in a laboratory environment and would look forward to the opportunity to assist in the day-to-day routine of a neurobiology lab. |
| In the first paragraph, he makes it immediately clear to what the letter refers stating the purpose and name of position.<br>The applicant then lists his accomplishments and goals, and he explains why he is a good fit for the position. | In May 2017, I will complete my undergraduate studies at the University of California–San Francisco with a B.S. degree in biology. I have done volunteer work in a research laboratory at the same university, assisting in the preparation of buffers, solutions, and helping in the operation of an HPLC for more than a year. I have greatly enjoyed this experience and would like to build on and expand my knowledge in a laboratory setup by starting my professional career in this environment. It would be an honor to join your research group as a Technical Assistant. I believe that my education, talents, and acquired skills as an undergraduate volunteer will be valuable assets for your laboratory. |
| The letter ends positively and indicates what has been enclosed. | Please find enclosed my résumé and the contact information for my references. My transcripts and letters of recommendation will be arriving under separate cover. Thank you for your consideration. I look forward to hearing from you.<br>Sincerely yours,<br>John Smith<br>*Enclosures* |

Following is another example of a cover letter:

| **Example 16-5** | **Cover letter 2** |
|---|---|
| The letterhead provides the name and contact information of the sender. | *Jane Miller* *33 River Rd.*<br>*Middletown, NY 10940*<br>*Email Telephone number* |
| This applicant does not know the name of the hiring manager; she therefore addresses the letter to the general responsible entity. | Date<br>Name<br>Address<br>Dear Hiring Manager:<br>As a recent graduate of the University of Maine with a Bachelor of Science, I am actively seeking your consideration for open positions, and am excited to apply for the posting as Program Coordinator in Development. |

The applicant's introductory paragraph provides general context and names the position and purpose of the letter.

My education, work experience, and personality all make me an excellent candidate for the position. I majored in biology with a minor in sociology. Much of my coursework has been focused on writing about a variety of topics in many different styles—from persuasive to reporting—a skill that can be considered one of my biggest strengths and that I greatly enjoy. I believe that these attributes will be invaluable for a career in development, a field that has always held a great interest for me. Additionally, working for a foundation during two consecutive summers as an undergraduate volunteer has taught me to pay close attention to detail, be extremely organized, and be able to multitask.

In the next paragraph, the applicant gives relevant information about herself that is applicable and important to the position. She shows how she can contribute.

Before graduating with a Bachelor's degree, I was given a unique opportunity to expand my skills even further as a research assistant for the Emergency Department at the X School of Medicine. In this position, my interpersonal skills and professionalism proved highly important to achieve the desired overall goals for the benefit of the department and its patients. I believe that these skills will also be an invaluable addition to your office.

The third paragraph continues the theme of the second, highlighting the applicant's strengths that are of importance to the position.

I would greatly appreciate the opportunity to meet with you and learn more about the program coordinator position as well as give you the chance to learn more about me. I believe I can be an excellent addition to your team and unit.

The applicant ends on a positive note, asking for a chance to meet in person, which shows that she is truly interested in the job.

Thank you in advance for your consideration.

Best regards,

Jane Miller

## 16.3 PERSONAL STATEMENTS

When you apply to graduate school, often a "personal statement" is required in addition to your CV, test scores, transcripts, and other documentation. Committees sorting through large numbers of applications need some way to select candidates from the pool of applicants. A personal statement allows the admission committee to differentiate candidates. Because your essay can be the deciding factor on whether you are accepted or rejected by an institution or other organization, it is important to carefully prepare this section of your application.

### ➤ Collect as much information as possible to write an engaging statement

Typically, your personal statement will be read by people who serve on an admissions committee in the department to which you are applying. Carefully consider this audience. You want your statement to engage the readers and to set you apart from the rest of the stack. For most people, the big challenge is to find an angle to make their personal statement engaging and interesting. So spend time finding a good "hook" for your statement.

Because applying for graduate or professional school or postgraduate employment is usually highly competitive, collect as much information as possible about the particular school, program, faculty, or employer before you apply. Consider speaking to a graduate school advisor or other contact person to obtain further insight. Knowing the areas of specialty of the department will help you tailor your application materials. It shows that you have done your homework and are seriously interested in the position. This gives you a potential advantage over other applicants.

### ➤ Content and organization

- Discuss relevant personal, academic, or research experiences
- Discuss future goals and plans
- Address the topic or question

A personal statement generally falls into one of two categories:

1. A general, comprehensive personal statement, which is usually the type expected for medical or law school applications.
2. A response to specific questions, which you should carefully tailor to the questions being asked.

For the first type of statement, discuss any relevant personal, academic, or research experiences and indicate your future goals and plans. For the second type of statement, address the topic or questions by providing a carefully crafted response.

A few actual essay prompts are shown following:

- Describe a skill, experience, or passion that shows what you would contribute to XXX University and how this will add to campus life.
- Describe an important experience you have had in the past 4 years that exposed you to a person outside of your own social or cultural group. Tell how this experience influenced you.
- In this section of the application, compose a personal essay justifying why you would like to study XXX science. In your essay, please do

not to exceed two double-spaced printed pages using 12-point type. Explain how your attributes, experiences, or interests would allow you to make a distinctive contribution to XXX University and YYY profession.

- In 500 words or less, provide an overview about your career progress to date. Expand on your future career plans, and elaborate on your motivation for pursuing a career in XXX field.
- Describe an accomplishment of which you are proud, and explain how this relates to your character.

Applicants should describe their personal and academic skills, research interests, and professional goals. They should tell how this program is a good match for their interests. Statements should not exceed one to two single-spaced pages.

Regardless of the form of the personal statement, committees are interested in learning why they should select you for the job or program. To address their concern, you need to answer the following questions:

- Why you?
- Why this field?
- Why this school or employer?

To answer the first question, explain what is special about you and your experiences, what your academic interests are, what your career goals are, what your personal character is, what your skills and expertise are, and why you are better qualified than others.

For the second question, give specific reasons why you are interested in a particular topic and what particular skills and expertise you would bring to the prospective field of study or position. Discuss your personal and/or academic background, your research experiences, and why you plan to attend graduate school or work in a particular field.

To answer the third question, explain why you are applying to a particular school or job, and what specific addition and skill set you will bring to the organization. What about the department's curriculum structure or general approach to the field attract you as a student or employee? Tailor your essay to match the program to which you are applying.

### ➤ Show how your interests, skills, and experiences match the program

In your essay, indicate how your interests, skills, and experiences are a good fit for the program and department overall. If you are interested in working with a particular faculty member, consider contacting the faculty member directly. Inquire with the faculty whether there is an opening in the laboratory before you apply to the program. In your essay, you may then indicate any prior communication with this faculty member about a position.

### ➤ Include an introduction, a main section, and a concluding paragraph

The most common format for a personal statement consists of three parts:

*Introduction.* The introduction, especially the first sentence, is the most important part of your statement. Therefore, the first sentence should attract attention and be creative, exciting, and short.

*Main Section.* This section should include three to five paragraphs. Here, you should support the main statement made in the Introduction. Discuss your experiences, accomplishments, or any other evidence as well as future goals. You may also want to include a brief overview of your educational background.

*Conclusion.* End your personal statement with a conclusion. For example, you may want to lay out why you are a good match to the department.

### ➤ Do NOT "tell your story"

Consider that a personal statement is a professional document that relays your value as an applicant and potential future addition to a school or company. Do not use it as a chance to "tell your story"—your experiences are already listed in your CV/résumé. Moreover, do not personalize or dramatize the assignment. You should also not write about all your chronological experiences. Rather, select and write about the experiences that make you a valuable applicant.

Similarly, do not discuss how hard you will work, how you will enjoy your time at the department, and how badly you want the job. Committees assume that everyone wants that opportunity. Instead, focus on *why* you want to work so hard toward this goal.

---

**Example 16-6** **Portion of a personal statement**

I was born in 1995 in Jackson Hole, Wyoming, into an Italian family. I attended elementary and middle school there and graduated from ZZZ high school in 2013. I then applied and was accepted into Purdue University. When I initially started at Purdue, I had not decided on a major, but greatly enjoyed all my science classes. Conservation biology, a field that had been of interest to me for some time, finally convinced me that I wanted to major in this topic. During my junior year, I applied for an undergraduate research position with Professor X, an expert in conservation biology, to dive deeper into the topic of species conservation, and quickly learned that a graduate degree would be needed for a solid career in the biological field. I am therefore applying to ABC University to make this realization become a reality. Since ABC University is one of the top universities in the field of conservation, a PhD degree from the university will be essential for my career as I am planning to eventually work at the EPA. . . .

---

**Revised Example 16-6**

Growing up in Jackson Hole, Wyoming, I was frequently exposed to nature through afternoons spent outdoors, and through weekend trips and summer hikes in the surrounding national parks. From an early age, I took notice and tended to wildlife. I brought home and cared for all kinds of animals—baby bunnies that had lost their mother, lizards, birds that fell out of a nest, and one day even a live snake that had been hit by a car. I nurtured these creatures and wanted to learn everything about them. I looked up their life cycles, biosphere, even their respective Latin names. Beyond the creatures I could bring home, I learned about animal migration, hibernation, and predation, and expanded my interests to include plants and eventually, during my high school and college years, the threat of human influences on their lives. Soon, I volunteered at park services. . . . When I attended university, my passion for nature and conservation was stoked even further. I eventually majored in conservation biology, a field that I would love to pursue much further and develop into a full-fledged career. . . .

### ➤ Start early

Writing an effective personal statement takes time. Before you begin to write, read over the essay directions on the application materials very carefully and then follow them exactly. You have to consider the topic, plan your points, structure your argument, draft the essay, revise it, and write a final version. Leave sufficient time to write the best essay possible. Make sure that you also allow time for revisions. Ask others to read your essay and to give you critical and honest feedback. Revise your essay until you are satisfied with it.

### ➤ Pay attention to details, including length, tone, and style

Aside from other factors such as your major, research area, faculty associations, potential funding, and background, admission committees look for how well an applicant writes and constructs an essay that is relevant and informative. They also note how well an applicant attends to details and demonstrates critical thinking and abstract reasoning, how well the essay is organized, and if the level of detail is appropriate. Hence, construct your essay carefully, organize it well, and pay attention to details.

Usually, personal statements are between 500 and 1,000 words long. Do not exceed any word limit given, and answer all the questions being asked. The tone of the essay should be objective or moderate. Do not sound too casual or too formal. Show confidence. Although personal statements use first person (e.g., "I," "we," "my"), avoid overusing "I." Instead, alternate between "I" and other first person terms, such as "my" and "me." Use active voice and transition words, such as "however" and "in addition." Ensure also that you have included a title as well as your name and contact information on the first page.

## ➤ Examples of personal statements

Following are two examples of personal statements written for graduate school. The first is a general, comprehensive personal statement. The second is the response to a specific question.

**Example 16-7** **General essay for graduate school application**

Mónica I. Feliú-Mójer Application to graduate program

| | |
|---|---|
| The first paragraph introduces the applicant (Mónica) by providing relevant background information on her as a person. | Growing up in a rural area of a small town in northern Puerto Rico, I was surrounded by nature and have since been captivated by its wonders. Collecting stones and insects was my favorite hobby, and my parents gave me a microscope for my ninth birthday. This fueled my scientific interest, making me wonder how nature worked, how everything was formed, and what the biological basis for events that surrounded me was. At the age of 11, when my father was diagnosed with a mental disorder, my interest was guided toward understanding how the brain works. |
| Mónica continues to describe her interest in science throughout her primary and secondary school education. She does so in an engaging way, as part of a story about her growing up and becoming a young adult. | My high school academic credentials granted me early admission to the biology program in XXX University. There, two words would eventually change my life: "Try research." These were the words of my freshman year biology professor, as she handed me an application for a summer research program at YYY University, one of Puerto Rico's medical schools. There, I had my first research experience as a member of Dr. R.J.C.'s neurophysiology lab. After spending the summer there, my interest and commitment to research were evident and I was selected as part of the MBRS program. During the two years I spent as an undergraduate in Dr. R.J.C.'s lab, I studied the behavioral and functional effects of repeated cocaine administration on rats, focusing on the functional changes in the noradrenergic neurotransmitter system. I had the opportunity to present my projects in several scientific forums and meetings, giving me the unique chance to interact with my peers and become more actively involved in the neuroscience research community. My undergraduate research experience solidified my interest in neuroscience and my desire to do research, pushing the boundaries of knowledge in this fascinating and still mysterious field. |

In this paragraph, Mónica further expresses her special interest in the field of neuroscience. It is clear that she has sought out opportunities to engage and learn more about research in this field. Her passion for neuroscience is apparent.

Since that initial research experience, I have been able to investigate other areas of neuroscience and physiology. In a research internship in Dr. P.A.'s lab at ZZZ University, I worked in the cardiovascular physiology field, doing in vivo evaluations of a mouse model of congestive heart failure. As a member of Dr. P.A.'s lab, I could compare differences between clinical research and the basic science I had performed earlier on in Dr. R.J.C.'s lab. My experience at ZZZ provided me with the opportunity to explore a completely different field of research and the reassurance that basic science, specifically in neurobiology, is my passion.

The applicant describes here how her interest in neuroscience turned into a first job experience, which she continues to enjoy.
It is clear that she has acquired a number of practical and technical skills.

To strengthen my skills and gain experience in neuroscience, I moved to the United States in 2004, to work in Dr. S.M.'s lab at the Center for Learning. At the age of 20, leaving my parents, friends, and family and giving up my first language was not easy. However, my will and determination to continue advancing in the neuroscience field proved to be strong, making my first experience as a full-time researcher a very enjoyable and productive one. As a Research Assistant in Dr. S.M.'s lab, I have explored the molecular mechanisms of synaptic plasticity using diverse molecular, cellular biology, and biochemical techniques such as Western blot analysis, yeast two-hybrid assays, dissociated neuron culture and cell culture, recombinant DNA cloning, immunoprecipitation, and generation and analysis of transgenic mice. In addition, I am responsible for administrative duties such as ordering, safety, overseeing laboratory animal issues, and keeping a common stock of drugs for the lab. Recently, one of my projects was published in the *Journal of Neuroscience Methods*. As a member of the S.M. lab, I have gained a richer understanding of the realities, hardships, and beauties of science, and also of the importance of social sensitivity—the ability to work with people from different countries, cultures, and scientific styles. Here, as I strengthen my laboratory skills, I practice how to handle multiple tasks and responsibilities, work hard, commit to goals, and overcome disappointments.

| | |
|---|---|
| Here, Mónica shows her character by describing her life philosophy. | Being born, raised, and educated in Puerto Rico, where scientific opportunities and resources are scarce, along with having Spanish as my first language, has challenged me to become more capable in order to compete and succeed. As a scientist, I understand the importance of making the population aware and informed about the advances being made in neuroscience, a very prolific, interesting, and fast-growing field. As a minority, I know that specific cultures and ethnic groups have different cultural and linguistic needs. Research, its results, and its impact on specific communities need to be addressed in a sensitive and nonstereotyped way. Being able to overcome these cultural and linguistic barriers gives me the unique opportunity to be a bridge between my field and my community and to encourage the participation of underrepresented minorities. |
| In the last paragraph, Mónica describes her reasons for wanting to apply to graduate school in general and to this one in particular. She indicates that she has very defined future goals and has informed herself about the program and school to which she is applying. Her essay has been crafted carefully toward her program of interest. | Graduate school is the next step in my pursuit of a career in neuroscience. H University's Graduate Program in Neuroscience, as well as its comprehensive and multidisciplinary approach, fits my prospective of a graduate program that will expand my knowledge base, grant me access to the most relevant and provocative questions, and allow me to meet mentors and peers who challenge me. With a broad variety of research interests among the faculty and the chance to perform my research in collaboration with various laboratories, H University offers me the opportunity to train in a highly interactive and diverse scientific environment. The high-quality training that H offers me will help me achieve my long-term goal of becoming an independent researcher, performing insightful and cutting edge research, teaching others, and giving people the chance to pursue their passion, as I am doing.<br><br>*(With permission from Mónica I. Feliú-Mójer)* |

---

**Example 16-8** **Specific response to a question**

| | | |
|---|---|---|
| | Annie Little | Application to occupational therapy program |
| Specific question to be addressed | *In 500 words or less, describe your main reason or experience leading up to applying for the university's occupational therapy program.* | |

This candidate's first paragraph immediately jumps into the story, providing the setting in a vivid and personal way.

Mrs. S. came from a large Italian family. She was very active and made Sunday dinner for about twenty people every week until she started showing signs of dementia. Eventually she ended up at a nursing home, and this is where I met her. This nursing home was very familiar to me because my mother worked there as a social worker, my oldest sister worked as an activity leader, and my other sister worked as a receptionist. I worked briefly in the activity department and also volunteered in other departments within the facility. When I was a volunteer, I was asked to keep Mrs. S. company because she appeared much more anxious when she was alone. I also realized that she seemed to get more agitated when others tried to take care of her, and she seemed to want to do more things for herself.

Then the applicant (Annie) describes her approach, commitment, and rewarding enjoyment in being with a geriatric patient.

Mrs. S. enjoyed sharing her juice and cookies with me as we sat and talked. I noticed that she resisted the nurses less when she was involved in an activity so the next time I visited her, I brought a puzzle for us to do. I explained to her that I needed her assistance. She struggled but we worked on the puzzle together and when it was completed, I had it framed and placed it on her nightstand. After that I brought something for her to do every time I visited, and one time I brought the ingredients and she made cannoli for the staff.

This paragraph provides further evidence of Annie's commitment and approach. It is clear that she is passionately devoted to helping Mrs. S. and has formed a personal connection—the basis for being a great future professional.

After several weeks, she seemed to be less anxious and better adjusted. Unfortunately involving someone in activities doesn't cure dementia, but when given activities to help others, her illness seemed much easier to manage and she seemed more oriented to the staff and the environment. Completing the puzzle and making cannoli had involved following specific steps, and she was able to complete these tasks with some direction and reminders. We also made signs together to help her remember simple tasks of daily living and labeled each of her drawers with words and pictures of what each contained.

The essay ends with a strong conclusion explaining why Annie wants to enter the occupational therapy field and states what her future goals are. It is within the 500 word limit (469 words total).

I feel that Mrs. S. had as much impact on me as I did on her because she made me realize that we should not take away a person's ability to do things for themselves. I saw how this frustrated her and made her anxious and agitated. Through my work with her, I realized that my passion lies with helping others and I believe work in the field of occupational therapy allows me to help others increase their independence and quality of life. I was exposed to the field of geriatrics because of my involvement with this nursing home. I have always been comfortable around individuals with disabilities, and my experiences have made me want to seek a career as an occupational therapist.

## 16.4 THE HIRING PROCESS AND INTERVIEW

### ➤ Realize the importance of networking

Networking, that is, making contact and creating a group of acquaintances to form mutually beneficial relationships, has become an important aspect in most scientific fields. Many students overlook this social aspect of professional training and obtaining a career. Networking may not always come easily or naturally for everyone, but nothing will be gained if you do not at least try to put yourself out there.

Networking can be done in a variety of ways—for example, in person, online through platforms such as LinkedIn, through an alumni database at your Career Services office, at conferences/meetings, etc. Take advantage of following up with people you have met through a thank you note, an email message, or by asking if you can meet again for coffee/lunch to informally interview the person about their job and receive mentoring advice. Consider getting business cards if you currently hold a position of employment and will be networking at conferences or large gatherings. Exchange contact information. Networking can lead to mentoring relationships, research collaborations, and possibly future jobs.

### ➤ Be patient during the hiring process

After you have submitted your job application, the response time can vary widely. In some cases, particularly in industry, you may hear back within a few days. Other times, you may not receive an acknowledgement or response to your application until a few weeks later. The latter is especially true in academia, where the hiring process is notoriously slow. To ensure your application is actually delivered to the hiring manager, consider writing to the manager directly in addition to sending the standard online application.

The hiring committee often takes time to collect, review, and then select applicants for an interview. Do not pester the hiring manager by requesting updates every week. Let the process of hiring take its course. You will be contacted if they want you to come to an interview.

### ➤ Prepare for the interview

If you are invited for an interview, inquire how the meeting(s) will be structured. For most positions, you will likely go through a half a day to a day of meetings and interviews. Prepare yourself for this interview process by doing your homework on the organization to which you are applying. Find out about their mission, organization, president, finances, products, news and so forth. In addition, ask an instructor or peers to give you a mock interview. Ensure that you have appropriate, professional/business attire, which typically includes suit and tie for men and business dress for women.

### ➤ Understand the most common interview questions

You can prepare an answer for many commonly asked questions ahead of time. Some common interview questions are listed below.

1. **Tell me about yourself.**
This is usually one of the first questions you will receive in an interview. The best reply is to pitch (briefly) why you are the right person for the position. Do not tell the interviewer your entire employment or personal history. Rather, describe your expertise and important achievements or relevant specific interests or experiences the interviewer should know about, and explain how this expertise or interest relates to the particular job. Show enthusiasm.
2. **How familiar are you with our organization?**
Here, the interviewer wants to see whether you are informed about the organization to which you are applying and whether you care about it. Show that you have done your homework and that you understand the organization's goals, mission, structure, and so on. You also should know the names of the top two to three key people. Then, explain why you are personally drawn to this job.
3. **Why are you interested in this job?**
Organizations want to hire people who are enthusiastic about a job. You should be able to explain why you are passionate about the position. State why the role and the company are a great fit for you.
4. **What do you consider your greatest strengths?**
For this question, it is important to list your true strengths (not those the interviewer may want to hear) that are relevant and specific to the position. Study the job posting beforehand, particularly the specifications and requirements, to know what is needed for the job.
5. **What are your weaknesses?**
This question allows the interviewer to find any red flags as well as to assess your honesty and your knowledge about yourself. Mention something that is a challenge for you but state that you are taking steps to improve. Do not state that you are perfect and have no faults, or that you have an irreconcilable problem ("I can never make it into the office before 10 a.m."). Rather, pick a weakness that you can turn it into a positive. (For example, you only know basic applications in Microsoft Excel but are taking a class to improve your skills in that respect.)
6. **Where do you see yourself in 5 years?**
The interviewer wants to see if you have thought about a goal and if the position aligns or leads you to this goal. Consider where this position could take you when you answer this question. If you are not quite clear about your ultimate goal, state that you view the position as an important step in determining this goal.
7. **Where else are you applying?**
This question will let the interviewer determine whether you are serious about the profession and what the competition is. It is best to state that you are looking at other, similar positions for which comparable characteristics are required. These characteristics should relate to skills you possess.

8. **What are your salary expectations?**
   Although you should avoid talking about salary during your interview, you have to be prepared to answer this question should it come up. Do your homework. Educate yourself about the current salary ranges for this type of position. If the interviewer asks the question, give a wide salary range and mention that the salary will not be an issue. You could state your previous pay to help the interviewer know the scale for negotiation, but it is best if the employer tells you the range first. Finally, if asked, you could respond with another question: "In what range do you usually pay people with my background?"
9. **How soon can you start?**
   This question does not mean you got the job. The interviewer might just like to have this information to plan for the position. If you have to discuss a transition with your current employer, say so. If you can start right away, feel free to mention that.
10. **Do you have any questions for me?**
    Aside from being questioned by the hiring manager, an interview also affords you the opportunity to ask questions in order to find out more about your potential future employer. Do not start your interview with your questions. Rather, wait until the opportunity arises naturally or until the interviewer gives you the chance. The next section discusses this aspect of the job interview in more detail.

### ➤ Think about your own questions ahead of time

Asking questions shows you are interested and also gives you the chance to garner enough information to decide whether the job is the right fit for you. The following are good questions for interviewees to ask, although I recommend restricting yourself to three to four questions total.

1. **What is a typical day or week like in the position?**
   The answer to this question will give you a good idea about what the position entails and how much time you will be expected to devote to certain tasks. The question will also show your interest in the position itself rather than just on the pay and benefits. One note of caution: Ask this question only if you are new to the position and do not have prior experience in the field.
2. **What are the biggest challenges for someone in this position?**
   This question allows you to be realistic about the job. Are you prepared to take on these challenges? The question also shows that you want to succeed.
3. **Is there room for professional development?**
   The answer to this question will allow you to gauge the hierarchy and upward movement within the organization. Are there clear guidelines and responsibilities for different levels, or is the position a static one? This question shows that you are ambitious and in for the long haul.
4. **How does your organization train employees for this position? Is there any mentorship?**

Asking these questions will help you find out if the organization cares about making their employees successful. Ideally, you do not want to be expected to pick up skills and knowledge simply by observing others. The question also shows that you are not just looking to do the bare minimum but rather to achieve in the role.

5. **How do you measure success in this position? What makes a person in this position outstanding?**
   The response will give you insight into what the manager values and about the metrics that will be applied. This question will also signal that you care not just about being average but about being great.
6. **Can you give me an overview of the organizational structure of the office?**
   The actual hierarchy of a place may not become clear until you receive a detailed organizational layout of the office or company structure. This will let you put the posted position into perspective and can also help you gauge internal movement and turnover.
7. **How would you describe the culture here?**
   Finding out about the overall environment of an office will let you determine if the position is the right fit for you.
8. **Could you describe your management style?**
   The answer to this question may provide some insight into what your manager expects and how he or she treats people. The answer may also raise a red flag about a difficult manager or unrealistic expectations.
9. **Are there any reservations you have about my fit for the position?**
   This question will allow you to address any doubts the hiring manager may have about you. The reply will also let you determine if these doubts are grounded and if the position might be a bad fit.
10. **What is the time frame for filling the position?**
    End your interview with this question so you will be clear about the next steps and the projected time frame.

## 16.5 LETTERS OF RECOMMENDATION

Along with your CV, cover letter, and personal statement (if requested), you will probably need one or more letters of recommendation to secure a good position. When you ask someone for a letter of recommendation, explain exactly why the letter is needed and how important it is to you. Always offer to provide information that makes the writing task easier (CV, list of accomplishments, publication list, the due date, means of transmission [mail or email], correct address of recipient). Make fulfilling this request as easy as possible for the letter writer. Include any instructions and information on the person or organization that the letter is supposed to go to. Explain to the recommender *why* this opportunity is important to you and *how* it fits into your overall career plan. If the writer cannot or will not provide you with a letter, accept this decline gracefully.

**Example 16-9 Sample letter of recommendation**

Susanna Turner
Office of Development
XYZ University
12 Ring Drive
Town, ST 09997

To
Embassy of Japan
ABC Program Office
2520 Massachusetts Avenue, N.W.
Washington, DC 20008
November 7, 2018

Dear Ms. Tanaka:

It is my highest pleasure to write this letter of recommendation for Kendall Nyef. Kendall has been working as a student worker in our office at the XYZ University Development office since fall 2016, where she has been assigned to support the office in general but has also worked with me personally on international initiatives of the university.

Despite her many other commitments as a full time student, Kendall has proven herself to be highly reliable, dedicated, and hard-working—if needed, even outside her regularly scheduled times. Her work ethic has been admirable. She has worked meticulously and diligently, and no matter what task—however small or big, laborious or fun, from creating file labels to analyzing donor lists—Kendall has more than exceeded our expectations. She usually went above and beyond the task at hand, making suggestions and improving processes and approaches. If needed, she did not hesitate to ask for further instructions. Her attitude has always been very positive.

Kendall would be particularly well-suited for the ABC Program. She is very personable, polite, and a joy to spend time with both at and outside of work. Her manner allows her to interact successfully on a professional level with colleagues as well as more informally with children. Additionally, her study abroad in Japan and travel to other foreign places has prepared Kendall for the challenges that living and working in a foreign country bring. She would be an ideal candidate for fostering mutual understanding between citizens of different nations. In short, Kendall has my strongest recommendation. I have no doubt that she will make an exceptional contribution to the ABC Program's mission of cultural exchange.

Please do not hesitate to contact me if I can provide further information.

With best regards,

**➤ Ask someone who knows you well, and allow plenty of time**
Approach someone who knows you well enough to include details about you as a person in the letter (an important reason as to why it is imperative to establish good relationships with your professors, supervisors, and peers). The writer should ideally write well, have experience composing letters of recommendation, and have the highest or most relevant job title. Also, plan your request. As a general rule, request your letter at least a month in advance. Some recommenders who have very busy schedules may even ask you to write a first draft yourself, which they will then use as a basis for their letter of recommendation. Do not hesitate to send a gentle reminder in email form to a busy professor to remind him or her about your request. You may also consider asking the recommender to send you an email confirming submission of the letter of recommendation.

## 16.6 CHECKLIST FOR A JOB APPLICATION

### OVERALL

- ☐ Did you tailor your CV or résumé specifically to the position and to the organization?
- ☐ Is your CV well presented and flawless?
- ☐ Did you tailor the cover letter specifically to the position?
- ☐ Did you ask your references before you provided their name and contact information?

### PERSONAL STATEMENT

- ☐ Is your statement engaging?
- ☐ Did you address the question or topic?
- ☐ Did you discuss your future goals and plans?
- ☐. Did you discuss how your interests match the department and faculty?
- ☐ Did you show what you can offer (e.g., hard work, research skills)?
- ☐ Did you discuss relevant personal, academic, and research experiences?
- ☐ Did you get feedback on the content?
- ☐ Did you stick to the word limit (i.e., one to two pages in length)?
- ☐ Did you use positive language?
- ☐ Did you proofread your essay?

### HIRING PROCESS AND INTERVIEW

- ☐ Did you re-read the job ad prior to your interview?
- ☐ Did you research the company and the company's website, become familiar with the most recent news articles, leadership, and the responsibilities of employees at this company?

- ☐ Did you research typical salaries for this position or for a comparable position?
- ☐ Did you prepare answers to the most commonly asked interview questions?
- ☐ Did you come up with a list of questions you would like to ask during the interview?
- ☐ Have you selected appropriate interview attire?

#### LETTER OF RECOMMENDATION

- ☐ Did you provide sufficient information about yourself and the job posting?
- ☐ Did you leave plenty of time for the recommender to compose a letter for you?
- ☐ Did you provide your recommender with detailed instructions for submission of the letter and the exact deadlines?

## SUMMARY

### JOB APPLICATION GUIDELINES

1. Tailor your CV or résumé specifically to the position and to the organization.
2. Highlight your academic achievements in your CV.
3. Highlight your skills and professional experiences in your résumé.
4. Ensure that your CV or résumé is well presented and flawless.
5. Format your CV or résumé strategically.
6. Customize the cover letter.
7. Know where to get help.

### PERSONAL STATEMENT GUIDELINES

1. Collect as much information as possible to write an engaging statement.
2. Discuss relevant personal, academic, or research experiences.
3. Discuss future goals and plans.
4. Address the topic or question.
5. Show how your interests, skills, and experiences match the program.
6. Include an introduction, a main section, and a concluding paragraph.
7. Do NOT "tell your story."
8. Start early.
9. Pay attention to details, including length, tone, and style.

### HIRING PROCESS AND INTERVIEW GUIDELINES

1. Realize the importance of networking.
2. Be patient during the hiring process.
3. Prepare for the interview.
4. Understand the most common interview questions.
5. Think about your own questions ahead of time.

### LETTER OF RECOMMENDATION GUIDELINES

1. Ask someone who knows you well.
2. Allow plenty of time.

## PROBLEMS

### PROBLEM 16-1

**Write your own CV or résumé. Get feedback on these documents from the career services office at your university or from faculty members.**

### PROBLEM 16-2

**Write a cover letter for a job or school to which you would like to apply. Get feedback on these documents from the career services office at your university or from faculty members.**

### PROBLEM 16-3

**Write a personal statement of less than 400 words using one of the writing prompts listed in Section 16.3. Have your essay evaluated in a peer group of four to five people, all of whom prepare similar statements for constructive feedback and group evaluation.**

# Appendix A

## COMMONLY CONFUSED AND MISUSED WORDS

The following list explains the meaning and use of the most commonly misused words scientific editors encounter. Strunk and White, Fowler, and Perelman have similar lists.

ABILITY, CAPABILITY, CAPACITY

| | |
|---|---|
| *ability* | The talent or skill to accomplish a specific thing. Ability can be measured; capacity cannot. (Some microorganisms have the *ability* to fix nitrogen.) |
| *capability* | The maximum, practical power or ability to do something; often relates to volume and quantities. (A tiger has the *capability* to carry prey of up to 550 kg.) |
| *capacity* | The maximum level or ability at which something can be held or contained, particularly in terms of volume. (The *capacity* of the beaker was 500 ml.) |

ACCEPT, EXCEPT

| | |
|---|---|
| *accept* | *Accept* means to answer affirmatively, to receive something offered with gladness, or to regard something as right or true. (The scientists *accepted* his new theory.) |
| *except* | *Except* is generally construed as a preposition meaning with the exclusion of or other than. It can also be a verb meaning to leave out. (Of the bacteria described, all are gram negative *except Streptococci.*) |

ACCURATE, PRECISE, REPRODUCIBLE

| | |
|---|---|
| *accurate* | *Accurate* means errorless or exact. It is often used in the sense of providing a correct reading or measurement. (The readings obtained with the spectrophotometer were *accurate.*) |
| *precise* | *Precise* means to conform strictly to rule or proper form. (The value 5.26 is more *precise* than the value 5.3.) |
| *reproducible* | *Reproducible* means something can be copied or repeated with the same results. (The experiment described by William et al. was *reproducible.*) |

ADAPT, ADOPT

*adapt* — *Adapt* is a verb and means to adjust and make suitable to. (Most organisms *adapt* easily to minor changes in the environment.)

*adopt* — *Adopt* is also a verb and means to accept or make one's own. (We *adopted* a new protocol for DNA isolation.)

ADMINISTER, ADMINISTRATE

*administer* — (The drug was *administered* orally.) *Administration* is the noun of *administer*, which may lead to this common confusion between *administrate* and *administer*.

*administrate* — *Administrate* means to manage or organize.

ADVICE, ADVISE

*advice* — *Advice* is a noun meaning suggestion or recommendation. (To measure *dP/dt*, we followed the *advice* of J. R. Boyd.)

*advise* — *Advise* is a verb and means to suggest or to give advice. (Dr. Boyd *advised* us to follow his protocol.)

AFFECT, EFFECT

*affect* — *Affect* is a verb and means to act on or influence. (The addition of Kl-3 to MZ1 cells *affected* their growth rate [i.e., it could have increased or decreased or induced].)

*Affect* can also be a noun with a specialized meaning in medicine and psychology: an emotion. (People can experience a positive or negative *affect* as a result of their thoughts.)

*effect* — As a noun, it means a result or resultant condition. (We examined the *effect* of Kl-3 on MZ1 cells.)

As a verb, it means to cause or bring about. (The addition of Kl-3 to MZ1 cells *effected* their growth rate [i.e., it had caused or brought about]. She was able to *effect* a change in his attitude.)

AGGRAVATE, IRRITATE

*aggravate* — *Aggravate* means to make something worse. (Bright light can *aggravate* a migraine.)

*irritate* — (pathologically) *Irritate* means to inflame, disturb, or cause pain or discomfort in a body part. (Certain chemicals can *irritate* the skin.)

ALLUDE, ELUDE, REFER, REFERENCE

*allude* — *Allude* means to mention indirectly. (The authors of "Basic Medical Microbiology" only *allude* to brain abscesses in Chapter 10.)

*elude* — *Elude* means to escape from or to escape the understanding or grasp of something. (The cause of the brain abscesses *eluded* us.)

*refer* — *Refer* means to pertain or to direct to a source. (*Refer* to Barett et al. for more detailed listings. Questions *referring* to brain abscesses should be directed to a specialist in the field.)

*reference* — *Reference* can be used as a noun, meaning a note in a publication referring the reader to another source (Do not forget to include *references* in the paper.)

The term can also be used as a verb, meaning refer to. (They *referenced* his work.)

ALTERNATELY, ALTERNATIVELY

*alternately* — *Alternate* means every other one in a series. (Students *alternately* attended lectures on molecular biology and biochemistry every Saturday.)

*alternatively* — A choice between two or more mutually exclusive possibilities. (Students can major in biology or, *alternatively*, in chemistry.)

AMONG, BETWEEN

*among* — *Among* means in a group of or the entire number of. It is used to express the relation of one thing to a group. (We discovered one black sheep *among* the white ones.)

*between* — Use *between* with two items or more than two items that are considered as distinct individuals. (We found no marked differences *between* our results and those reported in *Nature*.)

AMOUNT, CONCENTRATION, CONTENT, LEVEL, NUMBER

*amount* — Quantity that can be measured but not counted. (The total *amount* of yeast extract required for the medium was 12 g.)

*concentration* — The density of a solution or the amount of a specified substance in the unit amount of another substance. (The *concentration* of protein in the blood was 2.6 mg/ml.)

*content* — A portion of a specified substance. (Soybeans have a high protein *content*.)

*level* — Relative position or rank on a scale, often used as a general term for amount, concentration, or content. (Heart rate was at normal *level*. Protein *levels* [i.e., concentration]remained stable.)

*number* — A quantity that can be counted. (The *number* of proteins in the aggregates varied.)

ANYBODY, ANY BODY, ANYMORE, ANY MORE, ANYONE, ANY ONE

*anybody* — Refers to an unspecified person. Often used in place of everyone. (*Anybody* can get sick.)

| | |
|---|---|
| *any body* | A noun phrase referring to an arbitrary corpse or human form. (We found the head, but we did not see *any body*.) *This rule also applies to everybody, nobody, and somebody.* |
| *anymore* | An adverb denoting time. (Doctors in the Western world don't prescribe chloramphenicol *anymore*.) |
| *any more* | Used with a noun or as an indefinite pronoun. (We don't need *any more* measurements.) |
| *anyone* | Refers to any person. Used like anybody and everyone. (*Anyone* can get sick.) |
| *any one* | The two-word form is used to mean whatever one person or thing of a group. (I would like *any one* of these apples.) |

AS, LIKE

| | |
|---|---|
| *as* | *As* is a conjunction and is used before phrases and clauses. (The experiment was performed *as* described by Peters [3].) |
| *like* | *Like* is a preposition with the meaning of "in the same way as." (He was *like* a son to me.) Note the differences in the uses of *like* and *as*: *Let me speak to you as a father* (= I am your father and I am speaking to you in that character). *Let me speak to you like a father* (= I am not your father but I am speaking to you as your father might). |

ASSAY, ESSAY

| | |
|---|---|
| *assay* | A test to discover the quality of something. (In this *assay*, a Pt electrode was used.) |
| *essay* | A piece of writing, not poetry or a short story. (In this *essay*, she discussed her work in physical chemistry.) |

ASSUME, PRESUME

| | |
|---|---|
| *assume* | *Assume* means to take for granted or to suppose. It is usually associated with a hypothesis in scientific writing. (*Assuming* this hypothesis is correct, food should be supplemented with folic acid and vitamin B.) |
| *presume* | *Presume* means to believe without justification. (Scientists *presume* a common ancestor for the two species.) |

ASSURE, ENSURE, INSURE

| | |
|---|---|
| *assure* | *Assure* is to state positively and to give confidence and is used with reference to a person. (Let me *assure* you that we did not forget to add any enzyme.) |
| *ensure* | *Ensure* means to make certain. (When you ligate DNA, you have to *ensure* that you do not forget to add the enzyme.) |
| *insure* | *Insure* also means to make certain and is interchangeable with *ensure*. In American English, *insure* is widely used in the commercial sense of "to guarantee financially against risk." (We *insured* our car.) |

BECAUSE, SINCE

See SINCE, BECAUSE

BUT, AND, BECAUSE

*but, and,* and *because* — Can be used to start a sentence as long as the sentence is complete. (Some bacteria can form endospores. *But E. coli* and *E. sakazakii* do not.)

CAN, MAY

*can* — Means to be able to, to have the ability or capacity to do something. (Tetracycline *can* be used to treat urinary tract infections.)

*may* — Indicates a certain measure of likelihood or possibility to do something. It also refers to permission. (Tetracycline *may* act by binding to a certain site of the bacterial ribosome.)

CAN'T, DIDN'T, HAVEN'T

These contractions cannot be used in scientific writing at all. Write them out instead: *cannot, did not, have not.*

COMPLIMENT, COMPLEMENT

*compliment* — *Compliment* means praise. (John *complimented* Jean on her new dress.)

*complement* — *To complement* means to mutually complete each other. (The protein bindin and its receptor *complement* each other.)

COMPRISE, COMPOSE, CONSTITUTE

*comprise* — The conservative definition of *comprise* is to include or to contain. (The United States *comprises* many different states.) Avoid the phrase *is comprised of.*

*compose* — *Compose* means to make up or to create something. Frequently used in the passive voice. (Water *is composed of* hydrogen and oxygen.)

*constitute* — *Constitute* means to make a whole out of its parts, to equal or amount to. (Many different organisms *constitute* a habitat.)

CONSERVATIVE, CONSERVED

*conservative* — *Conservative* implies a medical treatment that is limited or treatment with well-established procedures. (Due to complications, we selected *conservative* treatment for this case.)

*conserved* — *Conserved* means to keep constant through physical or evolutionary changes. (DNA sequences may be *conserved* between species.)

CONTINUAL, CONTINUOUS

| | |
|---|---|
| *continual* | Repeatedly, occurring at repeated intervals, possibly with interruptions. (Measurements were hampered by *continual* interruptions.) |
| *continuous* | Without interruption, unbroken continuity. (Our experiments were *continuously* interrupted means the experiments were started and then interrupted, and this interruption never stopped. The light spectrum is *continuous*.) |

CONTRARY TO, ON THE CONTRARY, ON THE OTHER HAND, IN CONTRAST

| | |
|---|---|
| *contrary to* | Preposition meaning in opposition to. (*Contrary to* our expectations, addition of $Mg^{++}$ did not alter our results.) |
| *on the contrary* | *On the contrary* is used when one says a statement is not true. It is a subjective statement that indicates opposition and is usually used only in spoken English. ("It's exciting!" "*On the contrary*, it's boring!") |
| *on the other hand* | Use *on the other hand* when adding a new and different fact to a statement. Rarely used in scientific English. (It's cold, but *on the other hand*, it's not raining.) |
| *in contrast* | *In contrast* is also used for two different facts that are both true, but it points out the surprising difference between them. It is an objective statement for a marked difference and can be used in scientific writing. (pH values increased for procaryotes. *In contrast*, no pH difference was observed in eucaryotes.) |

DATUM, DATA

| | |
|---|---|
| *datum* | *Datum* is singular and means one result. |
| *data* | *Data* is the plural form of *datum*. It should be used with the plural form of the verb. (Our *data* indicate that many experiments have to be repeated.)<br>The same applies for *criterion*, *criteria*, and *medium*, *media*. |

DESCRIBE, REPORT

| | |
|---|---|
| *describe* | Patients and persons are *described*. (We *describe* a patient with gynecomastia induced by omeprazole.) |
| *report* | Cases and diseases are *reported*. (We *report* a case of omeprazole-induced gynecomastia.) |

DIE OF, DIE FROM

| | |
|---|---|
| *die of* | *Die of* is used for diseases and internal causes of death. (He *died* of cancer. She *died* of hunger.) |
| *die from* | *Die from* applies to external causes of death. (She *died from* her injuries. He *died from* drinking a poisonous substance.) |

DIFFERENT FROM, DIFFERENT THAN

*different from* Correct expression. (The binding site of amoxicillin is *different from* that of penicillin.)

*different than* Incorrect.

DOSE, DOSAGE

*dose* A *dose* is a specific amount administered at one time or the total quantity administered. (Patients received an initial *dose* of 10 mg.)

*dosage* *Dosage* is the rate of administering a dose. (The *dosage* he received was 500 mg every 6 hours.)

ENHANCE, INCREASE, AUGMENT

*enhance* Means to make greater in value or effectiveness.

*increase* A general term that means to become or to make greater. (Although we *increased* the amount of nutrients, the number of bacteria decreased.)

*augment* Means to make something already developed greater. (Addition of A greatly *augmented* the effect of B on C.)

In scientific writing, terms such as *increase* or *decrease* are preferred over *enhance* because they are much more precise.

ETC./SO ON, SO FORTH, AND THE LIKE

*etc.* Can only be used when the contents of a noninclusive list are obvious to the reader. It is an imprecise expression and should generally be avoided in scientific writing. Instead, use *such as* or *including* at the start of the list, and put nothing at the end of the list.

Also avoid *and so on*, *and so forth*, and *and the like.*

EXAMINE, EVALUATE

*examine* Patients are *examined.*

*evaluate* Conditions and diseases are *evaluated.*

FARTHER, FURTHER

*farther* Refers to a physical distance. (You need to drive much *farther* than you think.)

*further* Refers to quantity or time. (The binding of X to Z needs to be examined *further*.)

FEWER, LESS

*fewer* Use *fewer* for items that can be counted (*fewer* cells, *fewer* patients).

*less* Use *less* for items that cannot be counted (*less* medication, *less* water). Exceptions include time and money (*less* than two years ago; Jack has *less* money than John).

FOLLOW, OBSERVE

*follow* A case is *followed.*

*observe* A patient is *observed.*

To "follow up" on either approaches jargon, as does the term "follow-up study." However, in scientific writing, the use of the term *follow-up* is increasingly common, as is "follow-up study."

IMPLY, INFER

*imply* To *imply* means to hint at or suggest something. (These results *imply* that there is a key-lock mechanism between the enzyme and its receptor.)

*infer* To *infer* is to conclude or to deduce. (Looking at the interaction between the enzyme and its receptor, we can *infer* that a key-lock mechanism is used.)

INCIDENCE, PREVALENCE

*incidence* Means occurrence or frequency of occurrence. (The *incidence* of macular degeneration is high in the elderly.)

*prevalence* Means total number of cases of a disease in a given population at a specific time. (The *prevalence* of macular degeneration in the Western world in the year 2000 was 1 in 1,000 individuals.)

INCLUDE, CONSIST OF

*include* Means partial; to be made up of, at least in part; to contain. *Include* often implies partial listing. (Various antibiotics, *including* streptomycin and tetracycline, bind to the bacterial ribosome.)

*consist of* Means to be made up of, to be composed of. Usually used when a full list is provided. (Macrolites *consist of* . . .)

INFECT, INFEST

*infect* Endoparasites *infect* or produce an infection.

*infest* Ectoparasites *infest* or produce an infestation.

INTERVAL, PERIOD

*interval* The amount of time between two specified instants, events, or states. (Optical density was measured at 10-min *intervals.*)

*period* An interval of time characterized by certain conditions and events. (Pterodactylus lived during the Jurassic *period.*)

IRREGARDLESS, REGARDLESS

*irregardless* Does not exist in English. The correct form is *regardless.*

IT'S, ITS

*it's* Is a contraction of *it is* or *it has.* Preferably spell it out in scientific writing.

*its* Means belonging to. (Bindin is an adhesive protein found at the tip of the sperm cell. *Its* structure is unknown.)

LATER, LATTER, FORMER, LAST, LATEST

*later* Refers to time. (The fertilization membrane was not observed until *later* in the assay.)

*latter* Means the second of two items. (*Hpa*II and *Msp*I both cut the recognition sequence CCGG, the *latter* also when the sequence is methylated.)

*former* Means the first of two items mentioned. (*Hpa*II and *Msp*I both cut the recognition sequence CCGG, the *former* only when the sequence is not methylated.)

*last* *Last* refers to the last item of a series. (The samples were pooled, incubated at room temperature for 10 min, and centrifuged at 300 × *g* for 20 min. For the *last* step, 10 ml of 80% ethanol was added.)

*latest* *Latest* refers to the most recent item in a chronological series. (The *latest* book on molecular biology was published about one month ago.)

LAY, LIE

*lay* To place or put something. (*Lay* is a transitive verb; that is, it needs an object.) Past tense is *laid*; past participle is *laid.* (Birds *lay* eggs.)

*lie* To rest on a surface. (*Lie* is an intransitive verb; that is, it cannot take a direct object.) Past tense is *lay*; past participle is *lain*. (The tree line ends where the lake *lies.*)

LOCATE, LOCALIZE

*locate* To determine or specify the position of something. (Next, we *located* the vagus nerve.)

*localize* To confine or to restrict to a particular place. (YT-31 was *localized* in the mitochondria.)

MEDIUM, MEDIA

*medium* Singular noun. Needs a singular verb. (LB *medium* was prepared at room temperature.)

*media* Plural form of medium. Needs a plural verb. (For the refined analysis, we used six different types of *media*.)

MILLIMOLE, MILLIMOLAR, MILLIMOLAL

*millimole* Abbr. mmol; an amount, not a concentration. (A 1 *millimolar* solution contains 1 *millimole* of a solute in 1 liter of solution [or, a 1 *mM* solution contains 1 *mmol/liter* of solution].)

*millimolar* Abbr. mM; a concentration, not an amount. (A 0.5 *millimolal* solution contains 0.5 *mmol* of a solute in 100 g of solvent. The final volume may be more or less than 1 liter.)

*millimolal* A concentration, not an amount.

MUCUS, MUCOUS

*mucus* The noun. (*Mucus* is a viscous substance secreted by the *mucous* membranes.)

*mucous* The adjective meaning containing, producing, or secreting mucus.

MUTANT, MUTATION

*mutant* Refers to a strain of organism, population, allele, or gene that carry one or more mutations. (A *mutant* has no genetic locus, only a phenotype.)

*mutation* A *mutation* is an alteration in the primary sequence of DNA. (A *mutation* can be mapped, but a *mutant* cannot.)

NECESSITATE, REQUIRE

*necessitate* *Necessitate* means to make necessary or unavoidable. (The treatment may *necessitate* certain procedures.)

*require* *Require* means to need something. (She *required* medical attention.)

NEGATIVE, NORMAL

*negative* Tests for microorganisms and reactions may be *negative* or positive.

*normal* Observations, results, or findings are *normal* or *abnormal*. (The patient's behavior was *normal*.)

OPTIMAL, OPTIMUM

*optimal* Adjective meaning most favorable; never used as a noun. (The *optimal* [or *optimum*] temperature for X was 28°C.)

*optimum* Noun meaning the most favorable condition for growth and reproduction; often used as an adjective. (X reached its *optimum* 30 hours after induction.)

OVER, MORE THAN

| | |
|---|---|
| *over* | *Over* can be unclear. (The explosion was observed *over* 3 hours.) |
| *more than* | Use *more than* when you refer to numbers. (*More than* 500 people were treated.) |

PARAMETER, VARIABLE, CONSTANT

| | |
|---|---|
| *parameter* | Means a constant (of an equation) that varies in other settings of the same general form. Parameters are a set of measurable factors, such as temperature, that define a system and determine its behavior. (Changing the *parameters* of the system will result in a different outcome.) |
| *variable* | A *variable* is a quantity that can change in a given system. (In the equation $y = ax + b$, $x$ is the *variable*.) |
| *constant* | A *constant* is a quantity that is fixed. ($\pi$ is a *constant*.) |

PERCENT, PERCENTAGE

| | |
|---|---|
| *percent* | Means per hundred. Written as % together with number in scientific writing. (The yield of the gene product was less than 5%.) |
| *percentage* | Is a general part or a portion of the whole. Usually with a singular verb. (A small *percentage* of the crop was spoiled.) |

PERTINENT, RELEVANT

| | |
|---|---|
| *pertinent* | Means having logical, precise relevance to the matter at hand. (These assignments are *pertinent* to understand the class material.) |
| *relevant* | What relates to the matter or subject at hand. (He performed experiments *relevant* to his research.) |

PRINCIPAL, PRINCIPLE

| | |
|---|---|
| *principal* | *Principal* can be either a noun meaning a person or thing of importance or an adjective meaning main or dominant. (The *principal* reason for not observing any gas formation was low temperature.) |
| *principle* | *Principle* can only be a noun and means law or general truth. (This manual presents many writing *principles*.) |

QUANTIFY, QUANTITATE

| | |
|---|---|
| *quantify* | Often interchanged with *quantitate* and a matter of personal preference. Both are used with the meaning to determine or measure the quantity of something. |
| *quantitate* | *Quantitate* is preferred if one wants to emphasize that something was measured *precisely*. |

QUOTATION, QUOTE

*quotation* *Quotation* is a noun meaning a passage quoted. (Indicate a *quotation* with quotation marks.)

*quote* *Quote* is the verb and means to repeat or to cite. (Sentences copied from other sources should be *quoted*.)

RATIONAL, RATIONALE

*rational* An adjective meaning able to reason. (This is not a very *rational* thing to do.)

*rationale* A noun meaning basis or fundamental reason. (The *rationale* behind our theory is that cancer incidences are ever increasing.)

REGIME, REGIMEN

*regime* A *regime* is a form of government.

*regimen* *Regimen* means a regulated system or plan intended to achieve a beneficial effect. (He followed a strict *regimen* to lose weight.)

REMARKABLE, MARKED

*remarkable* Is often incorrectly used to indicate a change that is notable but not significant. The correct word is *marked*. (There was a *marked* increase in binding.)

*marked* See *remarkable*.

REPRESENT, BE

*represent* Means to stand for or to symbolize. (Each data point *represents* the average of five measurements.)

*be* Equal, constitute. (Mercury *is* a toxic substance.)

SINCE, BECAUSE

*since* Use this word only in its temporal sense and not as a substitute for *because*. (We have had nothing but trouble *since* we moved here.)

*because* If you want to indicate causality, use *because*. (The reaction rate dropped *because* temperature dropped.)

SYMPTOMS, SIGNS

*symptoms* *Symptoms* apply to people. (The patient displayed no *symptoms* of the disease.)

*signs* *Signs* apply to animals. (A *sign* that the dog was sick was that she did not eat anymore.)

THAN, THEN

| | |
|---|---|
| *than* | *Than* is a conjugation that introduces an unequal comparison. (Birds are closer related to dinosaurs *than* to mammals.) |
| *then* | Means next in time, space, or order. (The samples were centrifuged at 100 *x g*. *Then* the pellet was resuspended.) |

TOXICITY, TOXIC

| | |
|---|---|
| *toxicity* | *Toxicity* is the degree to which a substance is poisonous or toxic. (We determined the *toxicity* of compound X.) |
| *toxic* | *Toxic* means poisonous. (Mercury is *toxic* for most organisms.) |

USE, UTILIZE

| | |
|---|---|
| *use* | Generally, *use* is the preferred term. |
| *utilize* | *Utilize* means to find a new, profitable, or practical use for something. Viewed as an unnecessary and pretentious substitute for *use*. |

VARYING, VARIOUS

| | |
|---|---|
| *varying* | *Varying* means changing. (Precise measurements could not be taken because of *varying* levels of humidity.) |
| *various* | *Various* means different. (The crystals were of *various* sizes.) |

VIA, USING

| | |
|---|---|
| *via* | *Via* means by way of. (The students went to the lab *via* the ice cream parlor.) |
| *using* | *Using* means to put into service and by means of. (We isolated the plasmid DNA *using* a commercially available kit.) |

WHICH, THAT

| | |
|---|---|
| *which* | Use *which* with commas for nondefining (nonessential) phrases or clauses. (Dogs, *which* were treated, recovered.) |
| *that* | Use *that* without commas for essential phrases or clauses. A phrase or clause introduced by *that* cannot be omitted without changing the meaning of the sentence. Such essential material should not be set off with commas. (Dogs *that* were treated with antidote recovered.)<br>Sometimes the words can be used interchangeably. More often, they cannot. |

WHILE, WHEREAS, ALTHOUGH

| | |
|---|---|
| *while* | *While* indicates time and temporal relationship. It means "at the same time that." It is often incorrectly used instead of *although* or *whereas*. (Experiments were performed *while* patients were sleeping.) |

| | |
|---|---|
| *whereas* | *Whereas*, often the word the writer intended when using *while*, means "when in fact" and "in view of the fact that." (*Whereas* X increased, Y decreased.) |
| *although* | *Although* is a conjunction meaning despite the fact that or even though. (*Although* the association between breast cancer risk and variant alleles varied slightly, the *p* values were not statistically significant.) |

# Appendix B

## MICROSOFT WORD BASICS AND TOP 20 WORD TIPS

The following instructions apply to Word 2013 for Windows (PC version). On the Mac, most commands are the same but may have to be accessed differently.

WORD BASICS

In the screen display of the command area for Word 2013, the **Ribbon** displays the commands in task-oriented **groups** in a series of **tabs** (Figure B.1). Additional commands within the groups can be accessed by clicking on the **Dialog Box Launcher**. Selecting the **File Tab/ Backstage Button** in Word will bring you to a drop-down menu of commands such as *Save*, *Open*, *Close*, *Recent*, *New*, *Print*, and *Send*. You can customize the Quick Access Toolbar to contain buttons for tasks you perform frequently, such as *Save*, *Print Preview*, and *Undo Move*. The window size and screen resolution affect what you see in the command area of the program.

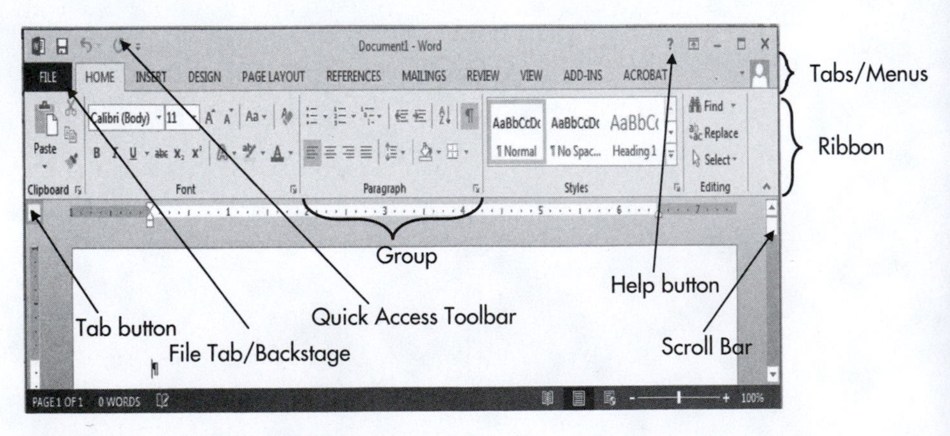


Figure B.1 Command area in Word 2013.

## TOP 20 WORD TIPS

**Tip 1.** If you do not know how to perform a specific function in Word, click on the **help button**, which is a white question mark in a blue round field on the top right corner of the command area, and type in the key word or question you have. The help function provides you with different options to narrow down your specific concerns and give you a step-by-step explanation to follow.

**Tip 2.** If you activate the **Show/Hide button (¶)**, which looks like a big π or pound sign in the Home tab Paragraph group, you can see the formatting symbols in your document as you work with it. This will enable you to identify spacing issues, including inconsistent spacing between words, hard page breaks versus sections breaks, and so on.

**Tip 3.** To apply the same formatting of an already formatted text section to another section, use the **Format Painter**, which can be found in the Home-Clipboard group. Highlight the source text, click the format painter button and then select the text to which you want to apply the formatting.

**Tip 4.** Use the **Print Preview** screen to see what your document looks like before you print it. To select Print Preview, click on the File Tab/Backstage button on the top left corner of the screen, then select Print and Print Preview automatically is displayed on the Microsoft Office Backstage view.

**Tip 5.** To write something as **super- or subscript**, highlight the letter or symbol you would like to convert to super- or subscript. Click on the applicable button in the Home-Font group ($\mathbf{x^2}$ for subscript, $\mathbf{x_2}$ for superscript).

**Tip 6.** If you need to add a special character, mathematical symbol, and/or foreign letter (such as α, β, or μ), choose the Insert tab and then the **Symbol** group. Select More Symbols and then the symbol/ character you want followed by Insert. If you go back to that same dialog box, Word displays the symbols you have most recently used.

**Tip 7.** To adjust **line spacing within paragraphs**, highlight the text for which you want to adjust the line spacing. Then, in the Home-Paragraph group, select the line spacing button and choose the line spacing option (1.0 for single, 2.0 for double, a specific value in between, etc.) that you would like.

**Tip 8.** To adjust **line spacing before and after paragraphs**, click anywhere within or highlight the paragraph and go to the Home-Paragraph group. Click on the lower right arrow and go to the Spacing section in the dialog box. Click on the Before and/or After arrows until you get to 0 pt for both, and then click OK.

**Tip 9.** To turn the **ruler on and off**, click on the View Ruler button on the right side just above the scroll bar, or select the Ruler box in the View Tab/Menu.

**Tip 10.** To **bullet or number text**, click within the paragraph that you would like to bullet or number and then select the bullet or numbering you want from the Bullets or Numbering button in the Home-Paragraph group. If the numbered list does not start where you want, click the small arrow next to the numbering button, select Set Numbering Value and choose the value you like.

**Tip 11.** Just below the main Ribbon on the left side is **Tab button**. This button designates the type of tab: left, center, right, decimal, etc. By clicking on the button, you can scroll through the different options. To set tabs for a particular section of text, first click in the line or paragraph you want to format. Then, scroll through the tab choices until you find the tab you want. Click on the horizontal ruler above the document at the point where you want to insert the tab, and the symbol for that tab will appear. You can click on the tab symbol on the ruler and slide it left or right as needed. You can use a combination of tabs to left-justify some text and right-justify the time at the end of the line. Using the right-justify tab is the only way to ensure that text on the right margin will be properly aligned there. (If you want to get rid of a tab, simply go to the text where the tab has been applied, click on the tab symbol on the ruler, and drag it off the ruler completely.)

**Tip 12.** To make a given **text section narrower or wider** than the normal page width in the rest of the document, highlight the applicable text and drag the left and/or right indents on the horizontal ruler above the page.

**Tip 13.** To **insert page numbers** that are automatically sequentially numbered, go to the Insert tab, chose Page Number within the Header & Footer group. The drop-down menu will allow you to select the placement and format of the page number. You can also designate which number to start with by selecting Format Page Number. If the formatting of the Word-inserted page numbers is not to your liking, you can simply change it as you would any text after double-clicking it.

**Tip 14.** To **insert a table** in Word, go to Insert Table and either highlight the rows and columns you need in the grid, or go to the Insert Table choice farther down in the drop-down menu, type the number of rows and columns you want, and click OK. (Note: There are advantages and disadvantages to using Word or Excel for tables. Word tables will give you more formatting options, but Excel tends to be easier to work with numbers, particularly where calculations are involved. You can do a simple sum in a Word table [in the Table Tools–Layout–Data group], using the Formula button.)

**Tip 15.** You copy a spreadsheet, chart, or picture from another Microsoft application (Excel, Visio, etc.) and paste it (under **Paste Special**) as an object into Word. You will be able to modify the object in your Word document, without going back to the original program.

**Tip 16.** To **crop a picture or object** in word, left-click the picture, which brings up the Drawing/Picture Tools-Format group. Click on Format, and then on Crop in the Size group. This will create a broken border around the picture. Position the cursor just tangent to one of the border segments, and then the cursor will change to a "handle" you can use to adjust the size of the picture as you see fit. Once you move the cursor away from the picture, the picture will remain the cropped size, and you can then enlarge it if necessary. (Note that the original, full picture/object is still available from the cropped version if you want to undo this.)

**Tip 17.** CTRL + Enter allows you to insert a **page break** at the place the cursor is located.

**Tip 18.** To **wrap text around pictures and objects**, click on the picture or object, and a Picture Tools toolbar will appear. In the Picture Tools–Format group, click on Text Wrapping and select Tight or Square.

**Tip 19.** **Tracked changes and comments** allow you to see edits and/or comments that others make to the document. You can then accept or reject each of the changes and respond to the comments as needed. To turn on tracked changes, go to the Review tab, and click on the Track Changes button in the Tracking group.

**Tip 20.** To **insert comments**, go to the Review tab, and click on the Insert Comment button in the Tracking group.

## WORD SPECIAL CHEAT SHEET

### Keyboard Shortcuts

| | |
|---|---|
| Three hyphens (—) + ENTER | Normal line across the page |
| Three underscore (___) + ENTER | Bold line across the page |
| Three equal signs (===) + ENTER | Double line across the page |
| Three hashes (###) + ENTER | Thick line with thin lines above and below across the page |

# Appendix C

## MICROSOFT EXCEL BASICS AND TOP 20 EXCEL TIPS

The following instructions apply to Excel 2013 for Windows (PC version). On the Mac, most commands are the same but may have to be accessed differently.

EXCEL BASICS

In the screen display of the command area for Excel 2013, the **Ribbon** displays the commands in task-oriented **groups** in a series of **tabs** (Figure C.1). Additional commands within the groups can be accessed by clicking on the **Dialog Box Launcher**. Clicking on the **File Tab/Backstage Button** in Excel 2013 is similar to clicking on that in Microsoft Word or PowerPoint. It displays a drop-down menu containing a number of options, such as *Open*, *Save*, *Send*, *Recent*, and *Print*. You can customize the Quick Access Toolbar to contain buttons for tasks you perform frequently, such as *Save*, *Print Preview*, and *Undo Move*. The window size and screen resolution affect what you see in the command area of the program.

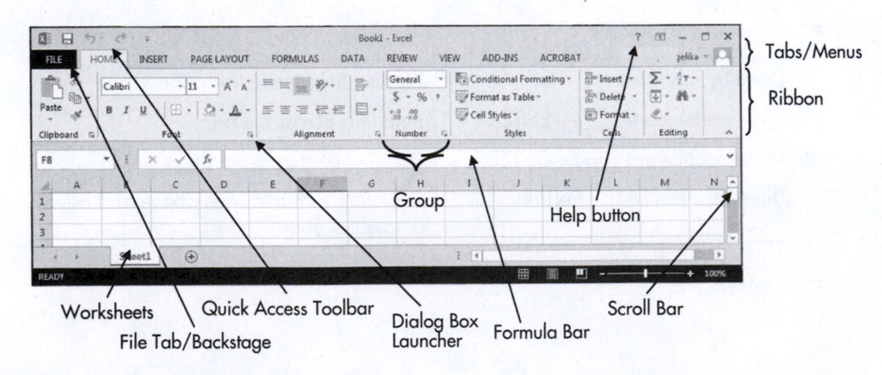


Figure C.1 Command area in Excel 2013.

## TOP 20 MS EXCEL TIPS

**Tip 1.** If you do not know how to perform specific functions in Excel, click on the **help button**, which is a white question mark in a blue round field on the top right corner of the command area, and type in the key word or question you have. The help function of Excel will provide you with different options to narrow down your specific concerns and give you a step-by-step explanation to follow.

**Tip 2.** To **navigate to a specific cell**, the arrow keys (▲, ▶, ▼, ◀) or the keys for PgUp and PgDn can be used instead of the mouse. Enter will go down one cell, and the Tab key will go to the next cell to the right. CTRL+Home will go to the first cell. CTRL+End will go to the last cell.

**Tip 3.** To apply the same formatting of one cell to another cell or to a cell on a different sheet, use the **Format Painter**, which can be found in the Clipboard group of the Home menu. Highlight the source cell(s), click the Format Painter button, and then select the cell(s) to which you want to apply the formatting.

**Tip 4.** Excel can **follow a sequence** you type in, such as months, years, or a simple series of numbers. For example, Excel can add in all the days of the week in subsequent cells once you enter Monday into the first cell. To create such a series, click on the cell where you typed in the first item in the series (Monday). Move your cursor over the small square (handle) at the right bottom corner of the selected cell, left-click on the square, and while holding down the mouse button, pull the handle over the cell(s) to which you would like to extend the series (down or right). Let go of the mouse button and the series will be displayed. This technique works for time, numbers, months, calendar, quarters, and a few other series. If you double-click the handle, Excel will fill in an entire series for a table column automatically.

**Tip 5.** To use a **formula or function** for ease in calculations and applications, start your cell input with an equal sign. For example, if you would like to multiply something by 2, write =x*2, where x can be a number or a cell (A1, B1, etc.). See below for specific formulas to use for diverse mathematical functions.

**Tip 6.** **CTRL+Roll Mouse wheel** zooms in and out within the Excel display window.

**Tip 7.** If you want to **add up items in a column and/or row**, click on the cell below the list. For instance, if you want to add the numbers in cells B1–B10, click your cursor into B11. Then select the Greek symbol Σ on the Home tab in the Editing group to display the sum of the numbers listed sequentially above the selected cell.

**Tip 8.** If you had a **number that you wanted to display** as a percentage, decimal, or dollar amount, highlight the cell, and on the Home tab, select the percent (%), decimal (.), or dollar ($) sign in the

Number group. Clicking on the corresponding symbol will display the number in the cell as a percent, decimal, or dollar amount.

**Tip 9.** For a **shortcut of the most common features** in Excel, place your pointer inside the cell(s), then click on the right mouse button. A menu with the most commonly used features will open. You can select the commands directly from there.

**Tip 10.** To **insert or delete an entire row or column**, select the column to the right of the place where you would like to insert a column, or the row below where you would like to insert a row. Right-click on the letter of the column or on the number of the row and then select Insert from the pop-up menu.

**Tip 11.** If you like to **use Windows Calculator** to perform quick calculations while in Excel, you can save time by adding it to your Quick Access Toolbar. To add a calculator, click Customize Quick Access Toolbar to the right of your Quick Access Toolbar, and then choose More Commands. On the left-hand side choose Commands Not in the Ribbon from the drop-down menu, and you'll see Calculator in the list on the left pane. Just click the Add button to add it to the toolbar.

**Tip 12.** To **place Excel data into Word**, you can:

a. Copy and paste the table from Excel into the Word file. This is the quickest and easiest way to move Excel data into a Word file. However, this method only lets you change values without relating to any corresponding formulas in other cells in the table.

b. Embed Excel tables into your Word document. Embedding the Excel data directly in Word enables you to change the data and related formulas in Word. Note, though, that the embedded table will no longer have a connection to the source file. Thus, there will not be any automatic updates to the inserted table when you change your source file, and vice versa. To embed Excel data into a Word document:

- Select and copy the range of data in Excel that you want to embed in your Word document.
- In the Word document, click at the place where you would like to insert the Excel data. Then, in the Clipboard group of the Home tab, choose the Paste command.
- Select a Paste option in the pull-down menu under the Paste command.

Once it is embedded, you can then click on the table in Word and edit the contents directly because it will display as an Excel spreadsheet. The only problem is that embedding a table in a Word file may make the file unwieldy as it increases file size substantially.

c. **Link Excel worksheets to a Word file**. If you link your Excel worksheet in Word to the Excel source file, the worksheet in

Word gets updated automatically whenever you modify your original Excel sheet. To link your Excel data, follow the exact same steps listed under (b), but select Paste Link from the Paste Special dialogue box instead of another paste option.

d. If you want to **insert an empty worksheet,** you can choose the Insert tab and then the Table group and there click the small arrow to choose Excel spreadsheet from the menu.

**Tip 13.** To **hide columns and rows** and present a sheet that focuses on just the work area, select the row or column header from the row or columns to be hidden. Right-click within the selected rows or columns and select Hide. To unhide the column or row, select the row or column before and after the hidden ones, right-click, and select Unhide.

**Tip 14.** To **show or hide formulas** when you are working in an Excel worksheet, you can alternate between viewing the values in the cells and displaying the formulas by pressing CTRL+` (grave accent mark). (Note: If you're having trouble finding the grave accent mark, it's on the same key as the "~" symbol.)

**Tip 15.** **To select a large data set**, select the top left-hand cell and then use CTRL+SHFT+↑/↓ to move to top/bottom cell in a column and row.

**Tip 16.** Use CTRL+↑/↓ to jump to top/bottom cell in a column.

**Tip 17.** When reading or inputting data into a large table, it is useful to **freeze the horizontal or vertical panes**, or both, to orient yourself and keep columns and rows visible while the rest of the worksheet scrolls up and down. To freeze horizontal panes, go to the row below it, select the View tab, and in the Windows group, click Freeze Panes. The row header will be "frozen" when you scroll downwards. To freeze vertical panes, go to the column on the right and repeat the same procedure. To freeze both vertical and horizontal panes, select the pane below and to the right, and repeat the procedure. To unfreeze panes, select *unfreeze* panes in the Windows group of the View tab.

**Tip 18.** To **turn on the gridlines** so that they will be printed, select the worksheet or worksheets that you want to print. On the Page Layout tab, in the Sheet Options group, select the Print check box under Gridlines.

**Tip 19.** You can **sort your data columns and rows** in various ways, from alphabetical to values and colors. To ensure that corresponding rows and columns stay together as such, be sure to click expand the selection after choosing custom sort from the sort and filter command in the Editing group of the home menu.

**Tip 20.** To **create a Chart** with the push of a keyboard button, select the data you want to include in the chart, including the labels. Press the F11 key or the ALT+F1 key.

## EXCEL SPECIAL CHEAT SHEETS

### Keyboard Shortcuts

| | |
|---|---|
| CTRL+' (apostrophe) | Copies contents of cell above |
| CTRL+Pg Up/Pg Dn | Move to next worksheet |
| CTRL+↑/↓ | Move to top/bottom cell in a column |
| CTRL+Shft+↑/↓ | Select to top/bottom of column |
| ALT+Enter | Wrap text in a cell |
| Click, then SHFT+Click | Selects cells between clicks |
| CTRL+Click – while holding the cursor over different cells | Selects all clicked items |
| CTRL+` (grave accent mark) | Shows formula for the cell |
| CTRL+; | Displays today's date |

### Fixing Common Error Messages

Error messages start with a pound (#) sign.

| | |
|---|---|
| ##### | If you see railroad tracks, your column is too narrow. (Solution: Widen the column.) |
| #REF! | Your formula refers to a cell that no longer exists, due to a change in the worksheet. |
| #DIV/0! | You are dividing by an empty cell or zero. (Solution: Fix the formula's denominator.) |
| #NUM! | A formula or function contains invalid numerical values. |
| #VALUE! | Your formula includes cells that contain different data types, such as text instead of a number. |
| #NAME? | Your formula contains text that Excel does not recognize, such as a type or missing punctuation. |
| CIRCULAR | When a formula refers to a cell in which it is located (e.g., A1+10 CANNOT BE USED in A1=10). |

### Formulas

| | |
|---|---|
| Addition of two cells | =A2+B3 |
| Addition of a constant | =A2+10 |
| Addition of a row of cells | =SUM(A1:C1) |
| Addition of a column of cells | =SUM(A1:A3) |
| Addition of a range of cells | =SUM(B1:C3) |
| Addition of scattered cells | =SUM(A2,B1,C3) |
| Subtraction of a constant | =A1–10 |
| Subtraction of a cell | =B2–B1 |

| | |
|---|---|
| Multiplication by a constant | =A1*10 |
| Multiplication of two cells | =A1*B3 |
| Multiplication by a percent | =A1*.40 or =B2*40% |
| Division by a constant | =A1/10 |
| Division by a cell | =A1/B2 |
| Exponentiation (square) | =A1^2 |
| Exponentiation (cube) | =A1^3 |
| Square root | =SQRT(A1) |
| Cube root | =A1^(1/3) |
| Increasing by a percentage | =A1+(A1*.4) or =A1*1.4 or =A1+(A1*40%) |
| Decreasing by a percentage | =A1-(A1*.06) or =A1*.94 or =A1-(A1*6%) |
| Calculate a percentage (Part/Sum) | =A1/$D$3 |
| Average of a column | =AVG(A1:A3) |
| Average of a row | =AVG(A1:C3) |
| Average of a range | =AVG(A1:B3) |

**Graphing with Excel**

**Tip 1.** To **create a customized chart** in Excel, select the entire data range (including the headings). Then, choose the Insert tab followed by the chart type from the Charts group. Choose the chart subtype.

**Tip 2.** **Changing the chart type** of your chart is easy. Right-click the chart, and then click Chart Type.

**Tip 3.** You can **change the chart size** by dragging its corners. You can also change its location by dragging it to the desired place.

**Tip 4.** You can **format text and numbers on a chart** by right-clicking them and using the mini toolbar.

**Tip 5.** To **format areas in a chart**, select them with a click and then use the Shape Fill group from the Format tab.

**Tip 6.** To **reshape an axis**, right-click on the axis and then choose Format Axis from the menu. Ensure that the axis options function is selected on the left pane. Choose Fixed on the minimum and maximum row, and type the axis starting and ending number. Then, select fixed on the Major unit row and enter the axis step between numbers. Click on Close.

**Tip 7.** To **plot a semilog graph** in Excel, double-click the *y* axis, then from the Scale tab select Logarithmic scale.

**Tip 8.** To **create a log–log graph** in Excel, you first need to create an XY (scatter) graph (see Tip 1 above). Select Logarithmic scale from the Scale tab.

**Tip 9.** Often, data tables contain empty cells or hidden data. It is possible to plot such data without a break in your plot and without first removing the empty cells or deleting hidden ones. Excel contains a function that allows you to choose **how empty cells and hidden data are plotted**. To configure such plots, click on the chart and select the Design tab under Chart Tools. Choose the Select Data button and click on Hidden and Empty Cells. This lets you select either to display empty cells as Gaps or Zero or to graph a continuous line by choosing the Connect Data Points with Line alternative.

**Tip 10.** To **add a trendline** or best-fit line to your graph, move the mouse cursor to *any data point* on the chart where you wish to create the trendline. When you left-click, Excel will highlight all the data points. Place the cursor on any of the highlighted points, then right-click and select Add Trendline from the pop-up window. Within the window that appears, select the display you want for your graph.

You can also display an equation and R-squared value on your graph by selecting Options in the window that pops up when you select Trendline, and then selecting Display Equation on Chart and Display R-squared Value on Chart.

# Appendix D

## MICROSOFT POWERPOINT BASICS AND TOP 20 POWERPOINT TIPS

### POWERPOINT BASICS

In the screen display of the command area for PowerPoint 2013, the **Ribbon** displays the commands in task-oriented **groups** in a series of **tabs** (Figure D.1). Additional commands within the groups can be accessed by clicking on the **Dialog Box Launcher**. Clicking on the **File Tab/Backstage Button** in PowerPoint displays a drop-down menu containing a number of options, such as *Open*, *Save*, *Send*, *Recent*, and *Print*. You can customize the Quick Access Toolbar to contain buttons for tasks you perform frequently, such as *Save*, *Print Preview*, and *Undo Move*. The window size and screen resolution affect what you see in the command area of the program.

The following instructions apply to PowerPoint for Windows (PC version). On the Mac, most commands are the same but may have to be accessed differently.

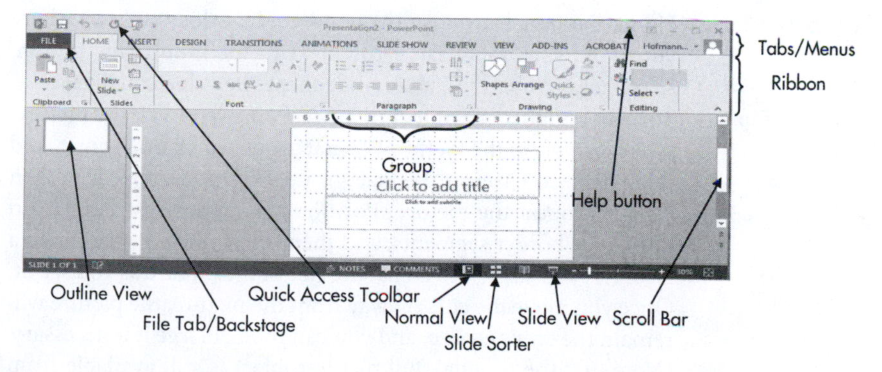


Figure D.1 Command area in PowerPoint 2013.

## TOP 20 POWERPOINT TIPS

**Tip 1.** If you do not know how to perform a specific function in PowerPoint, click on the **help button**, which is a white question mark in a blue round field on the top right corner of the command area, and type in the key word or question you have. The help function provides you with different options to narrow down your specific concerns and give you a step-by-step explanation to follow.

**Tip 2.** To apply the same formatting of an already formatted text section to another section, use the **Format Painter**, which can be found in the Home-Clipboard group. Highlight the source text, click the format painter button, and then select the text to which you want to apply the formatting.

**Tip 3.** If you need to **add a special character**, mathematical symbol, and/or foreign letter (such as α, β, or μ), choose the Insert tab and then the Symbol group. Select More Symbols and then the symbol/character you want followed by Insert. If you go back to that same dialog box, PowerPoint displays the symbols you have most recently used.

**Tip 4.** To adjust **line spacing within paragraphs**, highlight the text for which you want to adjust the line spacing. Then, in the Home-Paragraph group, select the line spacing button and choose the line spacing option (1.0 for single, 2.0 for double, a specific value in between, etc.) that you would like.

**Tip 5.** To adjust **line spacing before and after paragraphs**, click anywhere within or highlight the paragraph and go to the Home-Paragraph group. Click on the lower right arrow and go to the Spacing section in the dialog box. Click on the before and/or after arrows until you get to 0 pt for both, and then click OK.

**Tip 6.** To **bullet or number text**, click within the paragraph that you would like to bullet or number and then select the bullet or numbering you want from the Bullets or Numbering button in the Home-Paragraph group. If the numbered list does not start where you want, click the small arrow next to the numbering button, select Set Numbering Value, and choose the value you like.

**Tip 7.** To **crop a picture or object** in PowerPoint, left-click the picture, which brings up the Picture-Format group. Click on Format, and then click on Crop in the Size group. This will create a broken border around the picture. Position the cursor just tangent to one of the border segments, and then the curser will change to a "handle" you can use to adjust the size of the picture as you see fit. Once you move the cursor away from the picture, the picture will remain the cropped size, and you can then enlarge it if necessary. (Note that the original, full picture/object is still available from the cropped version if you want to undo this.)

**Tip 8.** To **turn on and off the ruler and gridlines** in order to help you to place text and objects in PowerPoint, right-click in the empty space on a slide and select ruler and/or grid and guides from the dialog box. Click OK.

**Tip 9.** If you want to **align two or more objects or text boxes** relative to each other, go to the Drawing group in the Home tab and select Arrange. In the dialog box, choose Align and then select the alignment you want (align left, align top, etc.)

**Tip 10.** To **insert a shape**, including a ring, bracket, or arrow, select the shape of choice from the Drawing group in the Home tab or from the Illustrations group under the Insert tab. Then, drag out the shape to the desired size. To modify the shape, click on the Format tab under drawing tools. Here you can change the shape fill color (or no fill), outline color, and shape effect. You can also adjust these under the Drawing group in the Home tab.

**Tip 11.** To **animate text and objects** in PowerPoint, go to the Animations tab and select the animation in the Animations group. An Animation pane dialog box will appear on the left side of the screen. To animate an object, select the object, and then select trigger in the Advanced Animation and timing and order in the Timing group. Fine tune the animation by right-clicking on the animation in the Animation pane and selecting Effect Options.

**Tip 12.** To **make text bullets appear one by one**, select the text in the Animation pane, then right-click and select Effect Options Text Animation Select by first-level paragraph.

**Tip 13.** To **hyperlink a slide** to another, or an object within a slide to another slide or file, right-click on the text or object, word or design element, then select Hyperlink from the dialog box. Choose Place in this document (or other address), then select the slide or file.

**Tip 14.** You can add diverse **Backgrounds**—such as solid- or gradient-color-designed background themes, and even imported pictures—to individual slides or to all slides in your presentation. There are two principal ways to add background to slides:

a. *From the Design Tab*
   - Under the Design tab of the ribbon, you can select various backgrounds in PowerPoint such as predesigned color schemes or custom background styles.
   - Clicking on Background Styles allows you to individualize colors, gradients, patterns, and more by selecting Format Background from the menu.

b. *Directly from the Slide*
   - When you right click next to or on a blank area of the slide, you can select Format Background from the shortcut window.
   - Select Fill to choose an individual background color and style.

**Tip 15.** Your can **add photos, illustrations, or documents** to your slides by simply cutting and pasting. You can also click on the Insert button in the ribbon and then select the image or illustration type you would like to add. Videos and Audio may also be added to PowerPoint. Images and illustrations in PowerPoint can be formatted using the Format tab of the Picture Toolbox.

**Tip 16.** To **resize an image without distorting** it, hold down the Shift key while grabbing and dragging the corner of the image. For illustrations or pictures that are too big for a slide, you may have to zoom out of the slide first (i.e., press and hold the control button while rolling the mouse wheel toward you).

**Tip 17.** To **change Views** in PowerPoint, you have several options (see also Figure D.1):

- Normal View is used primarily when you work on your slides.
- Slide Sorter View shows your slides as lined-up thumbnails, which you can view in different sizes by zooming in or out. This view allows you to rearrange your slides more easily by dragging and dropping them where needed.
- Slide View shows a full screen version of your slides as they are projected onto a screen.

**Tip 18.** You can **adjust the order of objects** in PowerPoint by selecting one of the commands in the Order Objects Menu within the Drawing-Arrange group in the Home tab.

**Tip 19.** To **print more than one slide per page**, select Print from the File Tab/Backstage Button, and then choose Handouts or Notes pages from the Print Settings options. Click Print.

**Tip 20.** To **paste text or an object or slide without losing its original layout and format**, paste the text or object into the slide, and then look for the Paste Options button that appears close to the inserted text, object, or slide. left-click this button and select Keep Source Formatting.

## POWERPOINT SPECIAL CHEAT SHEET

### Keyboard Shortcuts

| | |
|---|---|
| ESCAPE | Ends the presentation mode |
| CTRL+B; lowercase b; or full stop (.) during presentation | Blackens the screen during a presentation; click it again to continue |
| CTRL+W | Whites the screen during a presentation; click it again to continue |
| CTRL+H | Go to hidden slide |
| Holding SHIFT while dragging a shape | Allows you to drag the shape in only one direction to help in alignment |
| Holding SHIFT while drawing a shape | Makes shape perfectly symmetrical |

# Appendix E

## MICROSOFT OFFICE CHEAT SHEET

| | |
|---|---|
| F1 | Help |
| F4 | Repeats the last action |
| F12 | Save As |
| CTRL+Home | Beginning of document |
| CTRL+End | End of document |
| CTRL+Roll Mouse Wheel | Zoom in and out |
| CTRL+C | Copy |
| CTRL+X | Cut |
| CTRL+F | Find |
| CTRL+V | Paste |
| CTRL+P | Print |
| CTRL+S | Save |
| CTRL+Pg Up/Pg Dn | Move to next worksheet |
| CTRL+↑/↓ | Move to top/bottom cell in a column |
| CTRL+Shft+↑/↓ | Select to top/bottom of column |
| CTRL+Click—while holding cursor over components | Selects all clicked items |

# Answers to Problems

## Chapter 2 Fundamentals of Scientific Writing, Part I—Style

### Problem 2-1

1. Plants were kept [**at 0°C? or –20°C**] overnight.
2. [**10%, 12%, 6**] of exoplanets orbit multiple stars.
   POINT: "Some" is vague. How much is some? Use precise language.
3. The population of bivalve molluscs per square meter decreased markedly from 12 to three between 1999 and 2001.
   POINT: "Markedly" is vague. How much is markedly? Use a precise statement.

### Problem 2-2

1. support
2. used
3. are due to/are caused by different conditions.
4. To . . .
5. For example, . . .

### Problem 2-3

1. The doubling rate appeared to be short.
2. Homologous recombination is the preferred mechanism of DNA repair in yeast.
3. Often, jewel weed **grows close to** poison ivy.
4. Upon heat activation, filament size increased, and the number of buds decreased. Both **of these changes** were only seen for cytokinin mutants.

### Problem 2-4 Word Placement and Flow

Rainwater often picks up carbon dioxide, resulting in a weak solution of carbonic acid. When such rainwater trickles into the ground in areas with a high limestone content, the carbonic acid slowly dissolves the limestone, forming a cave. The cave grows underground as more and more limestone is dissolved. When a cave's ceiling gets eroded and collapses, a sinkhole forms.

### Problem 2-5 Word Placement and Flow

Fleas often carry organisms that cause diseases. An example of an organism that is transmitted by fleas to humans is the plague *bacillus.* These *bacilli* migrate from the bite site to the lymph nodes. Enlarged lymph nodes are called "buboes," giving rise to the name "bubonic plague."

### Problem 2-6 Word Placement and Flow

Thermophiles are microorganisms that grow at a temperature range between 45°C and 70°C. They are found in hot sulfur springs. Because thermophiles cannot grow at body temperature, they are not involved in infectious diseases of humans. However, it is unclear how they resist elevated temperatures.

*After reading the last sentence, the reader expects to hear about possible mechanisms of resistance.*

### Problem 2-7 Active Verbs

1. Transplant rejection **increased.**
2. Amyloid plaques **were measured** twice for each brain. (**We measured** . . .)
3. Our results showed that the vaccine **protected** the dogs.
4. To determine whether cells **migrate**, we dissected mouse brains.
5. Buffalo, elephant, and black rhino abundance all **declined** rapidly after 1977.

### Problem 2-8 Pronouns

1. A few microorganisms such as *Mycobacterium tuberculosis* are resistant to phagocytic digestion. **This resistance** is one reason why tuberculosis is difficult to cure.
2. Our findings indicate that binding decreases about 10-fold when temperature increases from 15°C to 25°C (Fig. 1). **This temperature increase/decrease/change** suggests that binding among the particles is not due to ionic interactions.

3. The color was achieved using new methods and concepts developed in our laboratory to distinguish specific cell wall components. **This coloration** in turn was only possible through heat induction.

### Problem 2-9 Parallelism and Comparisons

1. The pathogenesis observed in other cells, such as circulating monocytes, may differ from **that in** endothelial cells.
2. Diabetes can be affected **by both** exercise and diet. OR Diabetes can be affected **both by** exercise and **by** diet.
3. **Mutant A showed the highest peak.**
4. The male dolphin was **larger than** any of the other animals.
5. Many wasps can sting more often **than** honey bees.

## Chapter 3 Fundamentals of Scientific Writing, Part II—Compositions

### Problem 3-1

**(1′) Salmon use different methods to find their way during their homeward journey in the fall. (1)** Salmon use geomagnetic imprinting to return to their freshwater birthplace to spawn. **(2)** As described by Stabell et al. (15), **adults also use olfactory cues to guide them** back to the stream of their birth. **(3) It is unclear whether** salmon also use cues other than geomagnetic and chemical imprinting to orient themselves.

*The topic* (1′) *sentence states the message of the paragraph, "the different methods salmon use to find their way during their homeward journey in the fall." The topic is the subject in every sentence, resulting in a consistent point of view throughout the paragraph. Sentence 2 is parallel to sentence 1.*

### Problem 3-2

*A possible way to write this paragraph is the following:*

*CANCER CELLS*

Cancer cells, which are malignant tumor cells, differ from normal cells in three ways: (1) they dedifferentiate—for example, ciliated cells in the bronchi lose their cilia; (2) they can metastasize, meaning they can travel to other parts of body; and (3) they divide rapidly, growing new tumors, because they do not stick to each other as firmly as normal cells do.

## Problem 3-3

The most common active ingredient in insect repellents is DEET (N,N-diethyl-metatoluamide). DEET was developed by the U.S. military in the 1940s and is easily absorbed into the skin. Although DEET is considered a safe product if used correctly, there have been a few reports of adverse side effects in humans (Osimitz and Grothaus, 1995; Sudakin and Trevathan, 2003). Heavy and frequent dermal exposure can lead to skin irritation and, in rare cases, death, especially in young children (Osimitz and Grothaus, 1995, Sudakin and Trevathan, 2003). In addition, neurological damage and death can result from ingestion (Osimitz and Grothaus, 1995; Osimitz and Murphy, 1997).

## Problem 3-4

Global warming has been a major concern over the past decades as temperatures have increased globally due to human causes. As a consequence, ocean levels have risen 15–20 cm in the past century. This rise is due to (a) increased temperature leading to expansion of water and thus to a rise in sea level; and (b) a reduction of glaciers and ice sheets, further adding to the rising sea level. Worse, polar ice melt increases average temperatures on Earth as less sunlight is reflected from the polar caps. Instead, rising oceans absorb more light and heat, accelerating temperature elevation globally. As a result of these processes, ocean levels are expected to rise another 13–94cm by 2100 (Haug, 1998).

### Problem 3-5

*The paragraph can be condensed by only stating the most important information and omitting all other irrelevant results.*

Our results indicate that only *B. rutabulum* displays a pronounced increase in oxygen production at temperatures above 25°C.

*(18 words)*

## Chapter 4 Literature Sources

### Problem 4-1

Ostracodes are small bivalved Crustacea that form an important component of deep-sea meiobenthic communities along with nematodes and copepods (10). Crustaceans are dense and diverse in the deep sea and are one of the most representative groups of whole deep-sea benthic community **(10, 11)**. Pedersen et al. **(12)** as well as Jackson et al. **(13, 14)** reported that ostracode species have a variety of habitat and ecology preferences (e.g., infaunal, epifaunal, scavenging, and detrital feeders), representing a wide range of deep-sea soft sediment niches. Furthermore, Ostracoda is the only commonly fossilized metazoan group in deep-sea sediments (**reference**). Thus, fossil ostracodes are considered to be generally representative of the broader benthic community. The distribution and abundance of deep-sea ostracode taxa in the North Atlantic Ocean are influenced by several factors, among them, temperature, oxygen, sediment flux, and food supply **(14, 15)**. Several paleoecological studies suggest that these factors influence deep-sea ecosystems over orbital and millennial timescales **(1, 16)**.

### Problem 4-2

Droughts can occur in almost all climates, although their onset, severity, and frequencies vary widely. Droughts may be little noticed, if at all, in arid areas but result in dramatic loss of crops and water shortage in others. Droughts result when it has not rained enough over an extended period of time, usually a season or more. High temperatures and wind can add to their onset and lengths. Because scientists and politicians disagree about drought characteristics, not much progress in drought management has been seen.

### Problem 4-3

The deadly H5N1 avian influenza virus could spark a global pandemic that could kill tens of millions once it acquires the ability to spread in humans. Since 1997, H5N1 has been reported among humans, from Asia to Africa and Europe, although no cases have been detected in the United States. The virus has so far been passed on to humans mainly through poultry, but only a few gene mutations may be needed to make it transmissible among humans. Monitoring global bird and human outbreaks as well as stockpiling H5N1 vaccine will be key in recognizing and combatting a potential pandemic.

### Problem 4-4

*Version A would be considered plagiarized. Here, sentences have simply been rearranged, and a few words have been substituted. Overall, Version A is too close to the original. In contrast, Version B is paraphrased. It not only differs very much from the original, it also quotes the original source.*

## Chapter 5 Basics of Statistical Analysis

### Problem 5-1

$n = 15$
Mean = 14.2
Variance = 3.2
Standard deviation = 1.8

### Problem 5-2

a) 34.15%
b) 13.55%
c) 81.85%

### Problem 5-3

0.3% of the time (8mg.ml-5mg/ml=3mg/ml is three standard deviations or 99.7% of the time the brewing process will run)

### Problem 5-4

2.3%; 15.85%

### Problem 5-5

a) No correction needed.
b) Compound X ran on average 11.5 cm/hr on the thin layer chromatography (n = 12; σ = 2).

### Problem 5-6

*The result can be worded in various ways. Three examples are shown below for both a) and b). Note that in each case, the statistical information is subordinated parenthetically.*

a) The difference between the oxygen production of leaf segments 5 cm from the light source and those 20cm from the light source was significant ($p = 0.05$).
On average, the oxygen production of leaf segments closer to the light source (5 cm) was 20% higher that of those farther from the light source (12 bubbles/min). This difference was statistically significantly ($p = 0.05$).
On average, the oxygen production of leaf segments closer to the light source (5 cm) was significantly higher (20%; $p = 0.05$) than that of leaf segments farther away (20cm) from the light source.

b) The difference between the oxygen production of leaf segments 5 cm from the light source and those 20cm from the light source was not significant ($p = 0.75$).
   On average, the oxygen production of leaf segments closer to the light source (5 cm) was 20% higher that of those farther from the light source (12 bubbles/min). However, this difference was not statistically significantly ($p = 0.75$).
   On average, the oxygen production of leaf segments closer to the light source (5 cm) was not significantly higher (20%; $p = 0.75$) than that of leaf segments farther away (20cm) from the light source.

## Chapter 6 Data, Figures, and Tables

### Problem 6-1

1. Table or figure/map, depending if actual numbers are important or trends are meant to be displayed.
2. Table and/or figure/map—Table would provide precise data, but a map may show more visually the location of occurrence.
3. Table or graphs. Do not include the photo of the vivarium as it does not directly affect your experimental outcome.
4. Show the photograph of the newly discovered fungus—it presents new evidence. The other photographs do not present new evidence and therefore should not be shown.
5. Both—a schematic will help to visualize what is being described in the text.
6. Both—a schematic will help to visualize what is being described in the text.

### Problem 6-2

*A figure of the average values per time reading would best present the data.*

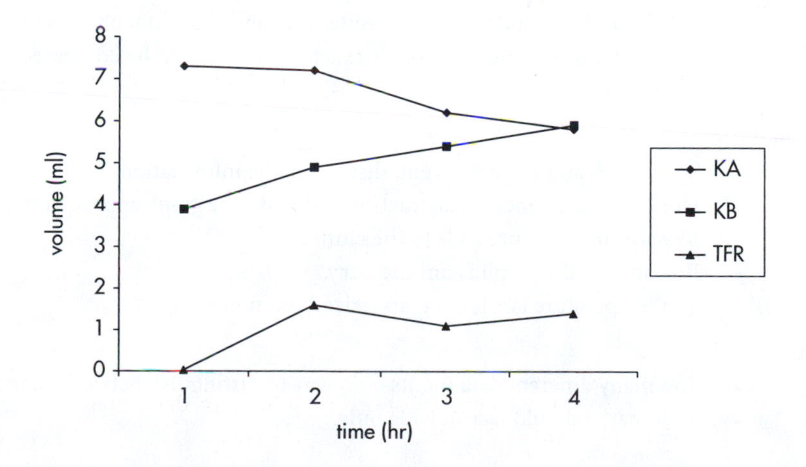

## Problem 6-3

*The best way to present this data would be as a bar graph.*

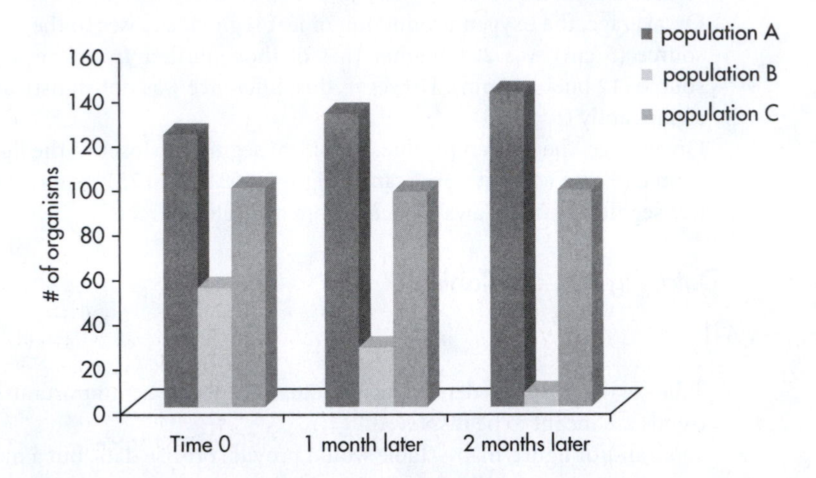


## Problem 6-4

- Omit the title—it should go into the figure legend.
- Expand the *y*-axis scale so data points do not fall below the *x*-axis.
- Omit the grid lines.
- Tick marks should only face outwards.
- Remove the heavy frame around the graph.
- Replace "+" symbols with closed circles, squares, or triangles.

## Problem 6-5

- Excessive numbering and tick marks on the *y* axis.
- Calibrations (tick marks) face inward where they are either unnecessary or may conflict with the data.
- Heavy black frame is distracting.
- Labels on the *x*- and *y*-axis are capitalized, and thus difficult to read.
- A line graph is plotted, although exact values seem to be compared and of importance.

## Problem 6-6

- Scales in A and B are different, distorting the information.
- Horizontal grid lines are distracting and make the graph appear cluttered.
- Key within the figure adds to the clutter.
- Box around the graph is unnecessary.
- The temperature labels are emphasized too much.

## Problem 6-7

- Too many different data points—no way to distinguish between them.
- Tick marks should face outside only.

- The number of scale marks on the $x$-axis should be increased and the spacing between them expanded.
- Scale marks on the $x$-axis should end at 2; those on the $y$-axis should end at 100.

## Chapter 7 Laboratory Reports and Research Papers

### Problem 7-1

| Section | Text |
|---|---|
| Background | To remove cellular metabolic wastes from the body, mammals have an excretory system. To gain more insight into such a |
| Question/Purpose | system, we dissected a kidney of a fetal pig. For this purpose, |
| Experimental approach | we cut from the lateral margin, leaving the ureter intact. |
| Main findings | We found that the kidney consists of three different internal regions: the outer cortex, the middle medulla (containing the renal columns and pyramids), and the inner renal pelvis. The kidney's nephrons filter water, nitrogenous wastes, and other materials from the blood, which enters the kidney through the renal artery above the renal vein. Water and waste material are collected and flow as urine through the collecting ducts into the renal pelvis. Ultimately, urine is then passed through the ureter, which empties into the urinary bladder. Filtered blood flows back to the body through the renal vein. |
| Conclusion | The study of the kidney shows how mammals eliminate their fluid waste. |

### Problem 7-2

*Despite good signals for the unknown and question, this introduction leaves the reader hanging and wondering what exactly was done and what the results are. The reason for its apparent incompleteness is that it does not state the question precisely nor mention any experimental approach. This introduction also provides a meaningless overview sentence of the discussion to come. Overview sentences such as these should be avoided in research papers, as they do not add anything to introductions of investigative papers.*

| Section | Text |
|---|---|
| Known | The carotenoid astaxanthin is a red pigment that occurs in specific algae, fish, crustaceans, and in some bird plumages (McGraw and Hardy, 2006). Astaxanthin is an antioxidant and commonly is used as a natural food supplement, food color, and anti-cancer agent among other disease preventative measures.<br>Like many carotenoids, astaxanthin is a colorful, fat/oil-soluble pigment, providing a reddish and pink coloration (2). Whereas in certain bird species all adult members display carotenoid containing feathers rich in color, many gulls and terns, which normally have white feathers, display an abnormal pink tinge (or flush) in various degrees across their populations (McGraw and Hardy, 2006). It has been |

| | |
|---|---|
| Known | suggested that this pink tinge arises during feather growth when these birds ingest abnormally high quantities of astaxanthin, which often occurs close to salmon farms (McGraw and Hardy, 2006). |
| Unknown | However, the exact relationship between astaxanthin and plumage is **not fully understood.** |
| Meaningless overview sentence<br>**Question/ Purpose—not stated precisely** | **Here we examine this relationship** in more detail and discuss its implication. |
| **Experimental Approach missing** | |

## Problem 7-3

*The original paragraphs contain many experimental details that should not be placed into the Results section. These details need to be removed as shown:*

To evaluate inhibitory effects of the isolated molecules, **inhibition assays were performed. Reaction products were analyzed by 12% PAGE and Western plot. We found that six out of 15 isolated molecules inhibited the kinase reaction at 10 μM markedly, and two of the molecules, A3 and A7, exhibited more than 50% inhibition of the enzyme activity. A 10-fold dilution series of these latter samples was prepared to determine the minimal inhibitory concentration. Molecule A3 exhibited 30% and molecule A7 45% inhibition of the kinase reaction at 1 μM. Buffer A did not interfere with the enzyme.**

## Problem 7-4

*The paragraph lists only results and conclusions but is missing background/ purpose information as well as the experimental approach:*

| | |
|---|---|
| **Purpose missing** | |
| **Experimental approach missing** | **To determine ... , we ...** |
| Results | We found that the H384A mutant reduced the $k_{cat}$ value more than 3-fold. The apparent $K_m$ values were increased 7-fold for Fru 6-P and 3.5-fold for PPi. |
| Interpretation | The increase of the $K_m$ values and the reduction of the $k_{cat}$ value of the H384A mutant suggest that the imidazole group of His384 is important for the binding stability as well as for catalytic efficiency of Fru 6-P and PPi substrates. |

## Problem 7-5

*Version B is a better first paragraph. It signals the answer, which is followed by supporting evidence. Version A, in contrast, summarizes and repeats results without interpreting them.*

## Problem 7-6

*Version A is a better concluding paragraph. The end is signaled clearly, the key findings are well summarized and interpreted, and the significance is indicated. Version B is simply listing limitations of the study, making for a very weak ending.*

# Chapter 8 Revising and Editing

## Problem 8-1

*Although some experimental approach, the results obtained, and the conclusion/implication of the work are stated, the abstract is incomplete. The author did not state the question or unknown/problem. The description of the results lets the reader guess what the question of the paper is, but it should not be the goal to have the reader guess. Another problem with this abstract is that no context is given for the experiments done. Furthermore, elements are not clearly signaled, nor are transitions used. A revised abstract is shown below. Note the in the revised version, context is provided, the unknown/problem is stated, and elements are clearly signaled.*

**Revised Problem 8-1**

| | |
|---|---|
| Background | **XXX disease has been associated with the bacterium *Enterobacter sakazakii*. However, the presence of this bacterium in infant milk formulas and in milk powder has never been investigated.** |
| Question/Purpose | **In this study,** |
| Experimental approach | **we used two different media (A and B) to test 50 powdered infant milk formulas and 25 milk powders for the presence of *Ent. sakazakii*.** |
| Results | **We detected** *Ent. sakazakii* in 5 out of 50 powdered infant milk formulas and in 4 out of 25 milk powders. Although both media detected *Ent. sakazakii*, medium A proved to be about twice as sensitive in the organism's detection than medium B. *Salmonella* bacteria, the standard organism used for testing food products for the presence of *Ent. sakazakii*, were not detected in either medium, |
| Conclusion/ Implication | **suggesting** that monitoring dry milk and food products solely for *Salmonella* is insufficient for the detection of *Ent. sakazakii*. |

# Chapter 11 Term Papers and Review Articles

## Problem 11-1

*The introduction of the review paper contains all necessary elements and is complete.*

| | |
|---|---|
| Background | Mitochondrial genomes differ greatly in size, structural organization, and expression both within and between the kingdoms of eukaryotic organisms. The mitochondrial genomes of higher plants are much larger (200–2400 kb) and more complex than those of animals (14–42 kb), fungi (18–176 kb), and plastids (120–200 kb) (Refs. 1–4). |
| Unknown/ Problem | Although there has been less molecular analysis of the plant mitochondrial genome structure in comparison with the equivalent animal or fungal genomes, |
| Topic statement | the use of a variety of approaches—such as pulsed-field gel electrophoresis (PFGE), moving pictures (movies) during electrophoresis, restriction digestion by rare-cutting enzymes, two-dimensional gel electrophoresis (2DE), and electron microscopy (EM)—has led to substantial recent progress. |
| Overview | Here, the implication of these new studies on the understanding of *in vivo* organization and replication of plant mitochondrial genomes is assessed. |

*(With permission from Elsevier)*

## Problem 11-2 Abstract

*The abstract is that of a research article. Sentences 1 and 2 provide a short background. Sentence 3 signals the question, whereas sentence 4 states the experimental approach. Results are signaled in sentence 5. Sentence 6 provides the conclusion by stating an implication. No topic statement or overview sentence is provided.*

| | |
|---|---|
| Background<br>Question/ Purpose<br>Experimental approach<br>Results<br>Answer/ Conclusion/ Implication | **(1)** Interleukin 1 (IL-1), a cytokine produced by macrophages and various other cell types, plays a major role in the immune response and in inflammatory reactions. **(2)** IL-1 has been shown to be cytotoxic for tumor cells (Ruggerio and Bagliono, 1987; Smith et al., 1990). **(3)** To determine the effect of macrophage-derived factors on epithelial tumor cells, **(4)** we cultured human colon carcinoma cells (T84) and an intestinal epithelial cell line (IEC 18) with purified human IL-1. **(5)** Microscopic and photometric analysis indicated that IL-1 has a cytotoxic effect on colon cancer cells as well as cytotoxic and growth inhibitory effects on intestinal epithelial cells. **(6)** Because IL-1 is known to be released during inflammatory reactions, this factor may not only kill tumor cells but also affect normal intestinal cells and may play a role in inflammatory intestinal diseases. |

## Problem 11-3 Abstract

*The abstract is that of a review article. Sentence 1 provides a short background. Sentence 2 gives an overview of the problem and together with sentence 3 describes the topic of the review. Sentence 4 provides an overview of the content.*

| | |
|---|---|
| Background<br>Problem<br>Topic<br>Overview of content | In the past few decades Africanized honey bees have been spreading throughout South and Central America into the southern part of the United States. During this spread, they have "Africanized" European bees largely through crossbreeding during mating flights, which almost always leads to the more aggressive Africanized bees. Various practices have been applied to counteract this trend. This review analyzes different practices used to counteract this trend and recommends the optimal approach to ensure pure European honey bees. |

## Problem 11-4 Review Title

*Both a) and b) would make good titles for review articles. Both are short, clear, and broad enough to be suitable for a review paper. Both also incorporate the top three to four key ideas/key terms of the paper. Example c) on the other hand, is quite specific, longer and a topic geared toward a research paper.*

### Problem 11-5 Conclusion

*The conclusion section nicely summarizes what the review presented. It starts with a general, brief overview of the topic, followed by a recap list of subsections covered. Finally, the conclusion section narrow down to a model that can be derived from comparison of sequences and the analysis of findings.*

### Problem 11-6 Conclusion

*The conclusion section is at best partial. It does not clearly summarize and generalizes all main lines of argument of the review. Although some of the items discussed in the review are summarized in the conclusion section, the conclusion overall is very short and ends with a vague overview sentence. This overview sentence leaves the reader hanging in terms of what was actually discussed in the review. Such a sentence could be expected at the end of the abstract or at the end of the introduction of a review article, but not at the end of the conclusion section, where actual, concrete conclusions, interpretation, and significance of the topic and results should be presented.*

### Problem 11-7 Main Analysis Section

a) *Outline for sea urchin species specificity:*

- Sea urchin species specificity of gametes
  - Intraspecies
  - Interspecies
    - Interspecies insemination frequency
- Surface macromolecules
    - For sperm
    - For egg
    - Macromolecule interaction
- Bindin deletion mutants
    - Species specificity
    - Comparison of protein sequences among different species
- Lock and key model

b) *Outline for snail hibernation:*

- Details of hibernation of land snails
  - Triggers:
    - Day length
    - Temperature
    - Water
  - Hibernation characteristics
    - Epiphragm formation
    - Feeding
    - Oxygen consumption
    - Heart rate
  - Evolution of hibernation characteristics in land snails
  - Comparison with other hibernators
    - Insects
    - Vertebrates

## Chapter 13 Oral Presentations

### Problem 13-1

*The slide shows a complex table taken directly from a published paper. Avoid using figures and tables directly as published in a paper, as most of the time you end up showing much more information than needed. If it is necessary to show a table, then reconstruct the table anew, and only show relevant portions on a slide. Avoid showing more than four columns and more than seven to eight rows. Listeners will not be able to absorb all this information at once, let alone read the small lettering of a reproduced figure or table. Also, consider highlighting values that are significantly different with a different colored font to draw the viewers' attention to the most important information on a slide.*

### Problem 13-2

*Several problems exist for this slide: (a) The figures have been copied directly from a publication; (b) they lack a title and unnecessarily show a figure legend; (c) the writing is too small to read and the copies look fuzzy; (d) the font of the heading is a serif font—it is better to use a sans serif font, such as Arial; (e) the heading is uninformative; and (f) the slide looks crowded.*

### Problem 13-3

1. *This ending is very weak and shows that the presenter is not very confident. It is better to end just with a simple "Thank you."*
2. *Asking the audience to focus only on part of a slide and ignore the rest tells the listeners that the presenter has not prepared his or her talk very well and/or that the slide is overcrowded.*
3. *The speaker here is using written English instead of spoken English. This sentence sounds very awkward when used in speech.*
4. *Use of the passive voice and written style is very heavy and boring to listeners. Instead, say, for example, "We determined that. . . ."*

### Problem 13-4

*Capturing the meaning of the slide by using as few words as possible is important. Details can be filled in by the presenter.*

**Value of Clinic**

- Welcoming environment
- Communication in native language
- Comprehensive care
- Professional and caring clinical teams
- Respected as human beings

## Chapter 14 Posters

### Problem 14-3

*The Abstract provided is too long for a poster and contains too many details. If an abstract is included on a poster, it should be only 50 to 100 words long. A shorter, more acceptable version is shown below.*

We studied the drought tolerance of brittlebush (*Encelia farinosa*) to understand when and how these bushes grow new roots and leaves and reestablish full metabolic activity. We found that a minimum of a half inch of rain is required for brittlebush to come out of dormancy. Our findings suggest that water needs to reach the deeper roots of the plants before dormancy is broken, and the longer the moisture is retained in the soil, the longer growth of the bushes is sustained.

### Problem 14-4

*The sample poster panel can be much condensed. Bullet points will also make the main information more readily available. A revised poster panel is shown following.*

**Conclusion**

- Recombinant bindin retains the ability to agglutinate eggs species-specifically.
- Species-specificity is contained in both amino- and carboxyl-terminal regions of *S. purpuratus* bindin.
- Repeated sequence elements in both amino- and carboxyl-terminal regions are different in bindins from different species.

## Chapter 15 Research Proposals

### Problem 15-2 Abstract

Cardiovirus is a common cause of gastroenteritis. For routine vaccination of Chinese infants, a new cardiovirus vaccine has been recommended. Here, we propose to evaluate the impact and cost-effectiveness of the Chinese cardiovirus vaccine program using a dynamic model of cardiovirus transmission. Findings from our analysis can be used to inform policy makers to prevent rotavirus infection.

### Problem 15-3

*This sample Abstract contains all necessary elements.*

---

**Abstract**

Background — The role that soils play in mediating global biogeochemical processes is a significant area of uncertainty in ecosystem ecology. One of the main reasons for this uncertainty is that

| | |
|---|---|
| Problem | we have a limited understanding of belowground microbial community structure and how this structure is linked to soil processes. Building upon established theory in soil microbial ecology and ecosystem ecology, we predict that the structure |
| hypothesis | of below-ground microbial communities will be a key driver of carbon and nutrient dynamics in terrestrial ecosystems. |
| Overall objective | We propose to test and develop the established theories |
| Strategy | by combining state-of-the-art DNA-based techniques for microbial community analysis together with stable isotope tracer techniques. By doing so, we expect to advance our |
| Significance | conceptual and practical understanding of the fundamental linkages between soil microbial community structure and ecosystem-level carbon and nutrient dynamics. |

*(Mark Bradford and Noah Frierer, proposal to private foundation)*

## Problem 15-4

*This Impact/Significance statement focuses on the scientist's needs and wishes rather than on the funder's or on society. Most funders will not be interested in the number of papers that will be published. They would like to see how your efforts will advance the field and their cause.*

***Possible revision:***

Funding from the Foundation will provide for a postdoctoral fellow and will lead to more insight into the fundamental principles of X, ultimately leading to better therapy and treatment options for Y.

## Problem 15-5

*The research proposal reads like an essay rather than a proposal. The text provides much background information and states the key questions to be answered but the overall objective is unclear. The objective is the most important statement of a proposal and needs to be included, even in boldface so the reader cannot miss it. In addition, in the text the specific aims are only alluded to but never stated outright. The largest section of the proposal, the experimental approach is completely absent. Thus, the reader comes away with interesting information about sea urchin gametes and their interactions but no idea about what the author is proposing to do and how.*

*A possible objective would be: To study the structure-function relationship of the sea urchin protein bindin and its receptor.*

*Specific aims could include:*

1. *Express bindin as a recombinant protein in E. coli.*
2. *Prepare and study the activity and species specificity of a series of deletion mutants of bindin.*
3. *Isolate and identify bindin's receptor.*

*Each of these aims should be followed by a detailed description of how the aim will be achieved experimentally and what expected outcomes would be.*

# Brief Glossary of Scientific and Technical Terms

**active verb** Verbs that express the action in a sentence (Example: "measure" instead of "measurement").

**APA style** Writing style format for academic journals and books, particularly in the social sciences, based on the style guide of the American Psychological Association (APA).

**biographical sketch** Brief description of a person's academic and professional accomplishments.

**BIOSIS** English-language, bibliographic database in the life sciences and biomedical sciences.

**Chicago style** Writing style format used largely in the social sciences and humanities based on the style guide for American English published by the University of Chicago Press since 1906.

**CINAHL** The Cumulative Index to Nursing and Allied Health Literature (CINAHL), an index of English-language and the largest nursing, allied health, biomedicine, and healthcare database.

**coherence** Logical relationship of sentences through linking of ideas from one sentence to the next.

**cohesion** Logical connection of sentences created through word location.

**conference abstract** Abstract of a presentation prepared of a conference in order to gauge interest in the presentation.

**continuity** The logical flow between sentences and/or paragraphs as a whole using all techniques of coherence and cohesion.

**core slide** Slide containing the most important information or figure in an oral presentation.

**CSE style** Writing style format created by Council of Scientific Editors (CSE; formerly CBE); used largely in the sciences.

**Current Contents** Search database that provides access to tables of contents, bibliographic information, and abstracts from the most recently published leading scholarly journals.

**curriculum vitae** Also known as CV; summary of a person's education, work experiences, and skills; used primarily in academic or medical settings.

**descriptive paper** Research articles describing a new discovery, apparatus, or application.

**EndNote** Program that helps you manage your references.

**ESL** Abbreviation for "English as a second language." Often referred to also as ESOL (English for speakers of other languages).

**funnel structure** Standard structure of the Introduction of a research paper, starting broadly with background information, narrowing to specific knowledge on a topic, to something unknown or problematic, and to the research question of the paper and its experimental approach.

**galley proofs** Semifinal form of a document that is laid out like the final form and provides last chance to make minor corrections before printing.

**Google Scholar** Free online database for academic publications.

**IMRAD** Format followed in writing a scientific research article; abbreviation stands for the order of the different sections of a scientific paper: Introduction, Materials and Methods, Results, and Discussion.

**indicative abstract** Abstract that provides the reader with a general idea of the contents of the paper; does not include any methods or results; used for review articles and book chapters.

**informative paper** Research article discussing research done in response to a hypothesis or to fill a specific gap.

**instructions for authors** Set of instructions and guidelines provided by scientific journals or funding agencies as a guideline for authors when writing a scientific article or proposal.

**investigative paper** Typical research article discussing research done in response to a hypothesis or to fill a specific gap.

**jargon** Use of terms specific to a technical or professional group.

**jumping word location** In consecutive sentences, the new information in the stress position of one sentence is placed at the topic position of the next sentence.

**key term** Words or short phrases used to identify important ideas in a sentence, a paragraph, and the paper as a whole; usually used to identify your main points in the topic sentence.

**key words** Important words that identify key ideas and points in a document; used for indexing and searching databases.

**LOI** Letter of intent, sometimes also referred to as "letter of introduction."

**MEDLINE** Database produced by the National Library of Medicine; broad coverage includes basic biomedical research and clinical sciences.

**MLA style** Modern Language Association (MLA), a commonly accepted writing style format, particularly in the liberal arts and humanities.

**NASA** National Aeronautics and Space Administration.

**NIH** National Institutes of Health.

**nominalizations** Abstract nouns derived from verbs and adjectives (Example: "measurement" instead of "to measure").

**noun clusters** A cluster of nouns or modifying words strung together, often incomprehensively.

**NSF** National Science Foundation.

**parallel form** Placing related ideas of equal weight into the same grammatical form and style.

**paraphrasing** To express someone else's words, thoughts, or ideas in your own words.

**peer review** Process of evaluating scholarly work (articles or proposals) by experts in the same field; also known as refereeing.

**plagiarism** Failing to indicate the source of information in scholarly scientific work or failing to paraphrase; a form of academic misconduct.

**power positions** Location where most important information in a sentence, paragraph, section, or document is placed. Two key positions exist: first and last, whereby first in a sentence, paragraph, or document is more powerful than last.

**preproposal** Short, overview version of a proposal; submitted ahead of proposal to gauge interest of a potential funder.

**presenter's triangle** Triangular space to the left side of the lectern; considered to best place to present without blocking the view of the audience to slides.

**primary sources of literature** Original, peer-reviewed publications of scientists' new research and theories.

**proposal** Grant application asking for financial support from a potential sponsoring agency.

**PsycINFO** Database that covers psychology and related disciplines.

**PubMed** A free database developed by the National Center for Biotechnology Information; it contains more than 22 million citations for biomedical literature from MEDLINE, life science journals, and online books.

**pyramid structure** Standard structure of the Discussion section of a research paper; starts with specific key results and their meaning and broadens to what these mean in the field generally and to society overall.

**résumé** Listing of your relevant job experiences and education; used primarily for industry job applications.

**review paper** Secondary sources representing a balanced summary of a timely subject with reference to the literature; they summarize what has been published or researched in a specific field and/or evaluate methods and results.

**RFP** Abbreviation for request for proposal, which is put out largely by funding agencies.

**SciFinder** Free online search database for scientific literature.

**SCOPUS** Database that provides broad international journal coverage of the sciences and social sciences.

**secondary sources** Literature source that cites, builds on, discusses, or generalizes primary sources (Example: a review article).

**subject/verb agreement** A singular subject has to be accompanied by a singular verb and a plural subject has to be accompanied by a plural verb.

**stress position** End position in a sentence where important, new information should be placed.

**tertiary sources** Literature source that generalizes and analyzes primary and secondary sources while attempting to provide a broad overview of a topic (Example: a textbook).

**topic position** Beginning position in a sentence where old, familiar information should be placed.

**topic sentence** First sentence of a paragraph, which provides an overview of the paragraph as well as important key terms.

**Web of Science** ISI Citation Databases that provide Web access to Science Citation Index Expanded for science and engineering journals.

# Bibliography


Alley, Michael. *The craft of scientific writing*, 3rd ed. Springer, New York, 1996.

Alley, Michael. *The craft of editing: A guide for managers, scientists and engineers*. Springer, New York, 2000.

Alley, Michael. *The craft of scientific presentations: Critical steps to succeed and critical errors to avoid*. Springer, New York, 2003.

Altman, Rick. *Why most PowerPoint presentations suck*, 1st ed. Harvest Books, 2007.

*American Medical Association manual of style*, 9th ed. Annette Flanigan et al., eds. Lippincott, Williams & Wilkins, 1997.

*American Medical Association manual of style: A guide for authors and editors*, 10th ed. Oxford University Press, New York, 2007.

American National Standards Institute, Inc. *American national standard for the abbreviation of titles of periodicals*. American National Standards Institute, New York, 1969.

American National Standards Institute, Inc. *American national standard for the preparation of scientific papers for written or oral presentation*. American National Standards Institute, New York, 1979.

American National Standards Institute, Inc. *American national standard for writing abstracts*. American National Standards Institute, New York, 1979.

Atkinson, Cliff. *Beyond bullet points: Using Microsoft Office PowerPoint 2007 to create presentations that inform, motivate, and inspire*. Microsoft Press, 2007.

Booth, V. *Communicating in science: Writing a scientific paper and speaking at scientific meetings*, 2nd ed. Cambridge University Press, Cambridge, 1993.

Browner, Warren S. *Publishing and presenting clinical research*. Lippincott, Williams & Wilkins, 1999.

Browning, Beverly. *Perfect phrases for writing grant proposals*, 1st ed. McGraw-Hill, 2007.

Burchfield, R. W. *Fowler's Modern English Dictionary*, 4th ed. Oxford University Press, 2004.

*Chicago Manual of Style*, 17th ed. University of Chicago Press, Chicago, 2017.

*The complete writing guide to NIH behavioral science grants*, 1st ed. Lawrence M. Scheier and William L. Dewey, eds. Oxford University Press, New York, 2007.

Covey, Franklin. *Style guide for business and technical communication*. Franklin Covey, 1997.

Davis, Martha. *Scientific papers and presentations*. Academic Press, London, 2002.

Davis, M. and Fry, G. *Scientific papers and presentations*, 2nd ed. Academic Press, London, 2004.

Day, Robert A. *How to write and publish a scientific paper*, 5th ed. Oryx Press, 1998.

Day, Robert A. and Gastel, B. *How to write and publish a scientific paper*, 6th ed. Greenwood, 2006.

Dodd, Janet S. *The ACS style guide: A manual for authors and editors*, 2nd ed. American Chemical Society, 1997.

Ebel, Hans F., Bliefert, Claus, and Russey, William E. *The art of scientific writing: From student report to professional publications in chemistry and related fields*, 2nd ed. Wiley-VCH, 2004.

Feibelman, Peter J. *A PhD is not enough: A guide to survival in science*. Basic Books, 1993.

Foley, Stephen Merriam and Gordon, Joseph Wayne. *Conventions and choices: A brief book of style and usage*. D.C. Heath, Toronto, 1986.

Fowler, H. W. *A dictionary of modern English usage*, 2nd ed. Oxford University Press, London, 1965.

Geever, Jane C. *The Foundation Center's guide to proposal writing*, 5th ed. Foundation Center, 2007.

Gerin, William. *Writing the NIH grant proposal: A step-by-step guide*. Sage Publications, Inc., 2006.

Gibaldi, Joseph. *MLA Handbook for writers of research papers*, 6th ed. Modern Language Association of America, 2004

Glatzer, Jenna. *Outwitting writer's block and other problems of the pen*. The Lyon's Press, 2003.

Gopen, George D. *Expectations: Teaching writing from the readers perspective*. Pearson Longman 2004.

Gopen, George D. and Swan, Judith A. The science of scientific writing. *American Scientist* 78 (Nov.–Dec. 1990): 550–558.

Greenbaum, Sidney. *The Oxford English grammar*, 1st ed. Oxford University Press, New York, 1996.

Gustavi, B. *How to write and illustrate a scientific paper*. Cambridge University Press, Cambridge, 2003.

Hacker, Diana. *Rules for writers*, 5th ed. Bedford/St Martin's, 2004.

Hall, Mary S. and Howlett, Susan. *Getting funded: The complete guide to writing grant proposals*, 4th ed. Continuing Education Press, 2003.

Harris, Dianne. *The complete guide to writing effective & award-winning grants: Step-by-step instruction*. Atlantic Publishing, 2008.

Henson, Kenneth. *Grant writing in higher education: A step-by-step guide*. Allyn & Bacon, 2003.

Huth, E. J. *How to write and publish papers in the medical sciences*. Williams & Wilkins, 1990.

Huth, E. J. *Writing and publishing in medicine*, 3rd ed. Lippincott, Williams & Wilkins, 1998.

Iles, Robert L. and Volkland, Debra. *Guidebook to better medical writing*, revised ed. Iles Publications, 2003.

*Illustrating science: Standards for publication*. CBE Scientific Illustration Committee. Council of Biology Editors, Bethesda, MD, 1988.

International Committee of Medical Journal Editors. Uniform requirements for manuscripts submitted to biomedical journals: Sample references. http://www.nlm.nih.gov/bsd/uniform_requirements.html, last accessed October 2012.

International Committee of Medical Journal Editors. Uniform requirements for manuscripts submitted to biomedical journals: Writing and editing of biomedical publication. http://www.icmje.org/, last accessed October 2012.

Katz, Michael Jay. *From research to manuscript*. Springer, 2006.

Knisely, Karin. *A student handbook for writing in biology*, 3rd ed. W. H. Freeman, 2009.

Knowles, Cynthia. *The first-time grant writers guide to success*. Corwin Press, 2002.

Korner, Ann M. *Guide to publishing a scientific paper*. Bioscript Press, 2004.

Lindsay, David. *A guide to scientific writing*, 2nd ed. Longman, 1995.

Lynch, Jack. *The English language: A user's guide*. Focus Publishing/R. Pullins, Newsburyport, MA, 2008.

Machi, Lawrence A. *The literature review: six steps to success*, 2nd ed. Corwin, 2012.

Malmfors, B., Grossman, M., and Garnsworth, P. *Writing and presenting scientific papers*. Nottingham University Press, Nottingham, 2004.

Matthews, Janice R., Bowen, John M., and Matthews, Robert W. *Successful scientific writing*. Cambridge University Press, Cambridge, New York, 1996.

McMillan, Victoria E. *Writing papers in the biological sciences*. Bedford Books of St. Martin's Press, Boston, 1997.

Morgan, Scott and Whitener, Barrett. *Speaking about science: A manual for creating clear presentations*, 1st ed. Cambridge University Press, Cambridge, New York, 2006.

Navidi, William. *Statistics for engineers and scientists*, 3rd ed. McGraw-Hill, 2010.

O'Connor Maeve. *Writing successfully in science*. Chapman and Hall, 1991.

Ogden, Thomas E. and Goldberg, Israel A. *Research proposals: A guide to success*, 3rd ed. Academic Press, 2002.

Peat, Jennifer and Barton, Belinda. *Medical statistics: A guide to data analysis and critical appraisal*, 1st ed. BMJ Books, 2005.

Peat, Jennifer, Elliott, Elizabeth, Baur, Louise, and Keena, Victoria. *Scientific writing: Easy when you know how*, 1st ed. BMJ Books, 2002.

Penrose, Ann M. and Katz, Stephen B. *Writing in the sciences: Exploring conventions of scientific discourse*, 2nd ed. Longman, 2004.

Perelman, Leslie C., Paradis, James, and Barrett, Edward. *The Mayfield handbook of technical and scientific writing*. Mayfield Publishing, 1997.

Pechenik, Ian A. *A short guide to writing about biology*, 7th ed. Longman, 2009.

Rogers, S. M. *Mastering scientific and medical writing: A self-help guide*. Springer, Berlin, 2006.

Rubens, Philip. *Science and technical writing: A manual of style*. Henry Holt, Inc., New York, 1992.

Rumsey, Deborah J. *Statistics II for dummies*. For Dummies, 2009.

Rumsey, Deborah J. *Statistics for dummies*, 2nd ed. For Dummies, 2011.

Ryckman, W.G. *What do you mean by that? The art of speaking and writing clearly*. Dow Jones-Irwin, 1980.

*Scientific style and format: The CBE Manual for authors, editors, and publishers*, 6th ed. Council of Biology Editors, 1994.

Sternberg, Robert. *The psychologist's companion: A guide to scientific writing for students and researchers*, 4th ed. Cambridge University Press, New York, 2003.

Strunk, W., Jr. and White, E. B. *The elements of style*, 3rd ed. Macmillan, New York, 1979.

Style Manual Committee, Council of Biology Editors. *Scientific style and format: The CBE manual for authors, editors, and publishers*, 6th ed. Cambridge University Press, 1994.

Sullivan, K. D. and Eggleston, Merilee. *The McGraw-Hill desk reference for editors, writers, and proofreaders*, 1st ed. McGraw-Hill, 2006.

Taylor, Robert B. *Clinician's guide to medical writing*, 1st ed. Springer, 2004.

Teitel, Martin. *"Thank you for submitting your proposal": A foundation director reveals what happens next*. Emerson & Church, 2006.

Thurman, Susan. *The everything grammar and style book*. Adams Media, Avon, MA, 2002.

Townend, John. *Practical statistics for environmental and biological scientists*. Wiley, 2002.

University of Chicago Press Staff. *The Chicago manual of style*, 15th ed. University of Chicago Press, 2003.

Venolia, Jan. *Rewrite right!: Your guide to perfectly polished prose*, 2nd ed. Ten Speed Press, 2000.

Venolia, Jan. *Write right! A desktop digest of punctuation, grammar, and style*, 4th ed. Ten Speed Press, 2001.

Wason, Sara D. *Webster's New World grant writing handbook*, 1st ed. Webster's New World, 2004.

Williams, Joseph M. *Style: Ten lessons in clarity and grace*. Scott, Foresman & Co. 1988.

Williams, Joseph M. *Style: Lessons in clarity and grace*, 9th ed. Longman, 2006.

Yang, Jen Tsi. *An outline of scientific writing. For researchers of English as a foreign language*. World Scientific Publishing, 1995.

Yang, Otto O. *Guide to effective grant writing: How to write a successful NIH grant application*. 1st ed. Springer, 2005.

Young, Petey. *Writing and presenting in English: The Rosetta stone of science*, 1st ed. Elsevier Science, 2006.

Zeiger, Mimi. *Essentials of writing biomedical research papers*, 2nd ed. McGraw-Hill, 2000.

# Credits

## SOURCE REFERENCES FOR EXAMPLES AND PROBLEMS

**Problem 4-1** *Proceedings of the National Academy of Sciences* 105(5). Abrupt climate change and collapse of deep-sea ecosystems, 1556–1560. Yasuhara, M., Cronin, T. M., deMenocal, P. B., Okahashi, H., and Linsley, B. K. Copyright 2008, National Academy of Sciences, U.S.A.

**Example 7-24** Reprinted from *Developmental Biology* 156(1). Lopez, A., Miraglia, S. J., and Glabe, C. G. Structure/function analysis of the sea urchin sperm adhesive protein bindin, 24–33. Copyright 1993, with permission from Elsevier.

**Problem 7-5** Reprinted from *Acta Tropica* 89(2). Yaw Asare Afrane, Eveline Klinkenberg, Pay Drechsel, Kofi Owusu-Daaku, Rolf Garms, and Thomas Kruppa. Does irrigated urban agriculture influence the transmission of malaria in the city of Kumasi, Ghana? 125–134. Copyright 2004, with permission from Elsevier.

**Example 9-1** With permission from Tammy Wu.

**Example 9-2** With permission from Amanda Miller.

**Example 11-6** Reprinted, with permission, from the *Annual Review of Ecology, Evolution and Systematic* 38. Aronson, R. B., Thatje, S., Clarke, A., Peck, L. S., Blake, D. B., Wilga, C. D., and Seibel, B. A. Climate change and invisibility of the Antarctic benthos, 129–154. Copyright © 2007 by Annual Reviews, www.annualreviews.org.

**Example 11-7** Reprinted from *Trends in Ecology & Evolution* 21/11. Kellogg, C. A., and Griffin, D. W. *Aerobiology and the global transport of desert dust*, 638–644. Copyright (2006), with permission from Elsevier.

**Example 11-8** Reprinted from *Biological Conservation* 128. Solberg, K. H., Bellemain, E. Drageset, O., Taberlet, P., and Swenson, J. E. *An evaluation of field and non-invasive genetic methods to estimate brown bear (Ursus arctos) population size*, 158–168. Copyright (2006), with permission from Elsevier.

**Problem 11-1** Reprinted from *Trends in Plant Science* 2(12). Backert, S., Nietsen, B. L., and Boerner, T. The mystery of the rings: Structure and replication of mitochondrial genomes from higher plants, 477. Copyright 1997, with permission from Elsevier.

**Example 12-1** With permission from Nora Chov, University of Connecticut.

**Example 12-8** With permission from Nikolas Franceschi Hofmann, University of Connecticut.

**Example 13-2a** With permission from Betty Liu, Yale University and currently Keck Foundation.

**Problem 13-1** Table reprinted from *Analytical Biochemistry* 230(1). Hofmann, A., Tai, M., Wong W., and Glabe, C. G. A sparse matrix screen to establish initial conditions for protein renaturation, 8. Copyright 1995, with permission from Elsevier.

**Problem 13-2** Figure reprinted from *Developmental Biology* 156(1). Lopez, A., Miraglia, S. J., and Glabe, C. G. Structure/function analysis of the sea urchin sperm adhesive protein bindin, 10. Copyright 1993, with permission from Elsevier.

**Example 14-2** With permission from Roland Geerken.

**Example 14-4** With permission from Roland Geerken.

**Example 14-5** With permission from Roland Geerken.

**Example 14-7** With permission from Philip Duffy, Yale University.

**Example 15-1** With permission from Jun Korenaga, Yale University.

**Problem 15-3** With permission from Mark Bradford, Yale University.

**Example 16-7** With permission from Mónica I. Feliú-Mójer, Harvard University.

## REFERENCES WITHIN EXAMPLES AND PROBLEMS

**Example 3-1: Revised Example 3-1a, b** Witham, CS. 2005. *J Volcanology and Geothermal Res* 141(3–4), 299–326.

**Example 3-12** Jones, BH et al. 1996. *Am J Physiol* 270, E192–E196.

**Problem 3-1** Stabell, OB. 1984. *Biol Rev*, 59(3), 333–388.

**Problem 3-3** Osimitz, TG and Grothaus, RH 1995. *J Am Mosquito Control Assoc* 11 (2 Pt 2), 274–278; Osimitz, TG and Murphy, JV. 1997. *Clin Toxicol* 35(5), 435–441; Sudakin, DL and Trevathan, WR. 2003. *Clin Toxicol* 41(6), 831–839.

**Problem 3-4** Haug, GH et al. 1998. *Paleoceanography*, 13, 427.

**Example 7-8** Chan, DK and Hudspeth, AJ. 2005. *Nature Neurosci* 8(2), 149–155; Brownell, WE et al. 1985. *Science* 227(4683), 194–196; Santos-Sacchi, J et al. 2006. *J Neurosci* 26(15), 3992–3998.

**Example 7-9** Vacquier, VD and Moy, GW. 1977. *Proc Natl Acad Sci USA* 74(6), 2456–2460; Bellet, NF et al. 1977. *Biochem Biophys Res Commun* 79 (1), 159–165.

**Example 7-23** Riffell, JA et al. 2008. *PNAS* 105(9), 3404–3409.

**Problem 7-2** McGraw, KJ and Hardy, LS. 2006. *J Field Ornithol* 77(1), 29–33.

**Example 10-1, 10-2** Schmittner, A. et al., 2011. *J. Climate* 24, 2814–2829.

**Example 11-8** Wiegand, T. et al., 2003. *Oikos* 100, 209–222; Taberlet, P. et al., 1996. *Nucl Acids Res* 24, 3189–3194; Hedmark, E. et al., 2004. *Conserv Genetics* 5, 405–410; Maudet, C. et al., 2004. *Mol Ecol Notes* 4, 772–775.

**Problem 11-2** Ruggerio, V and Baglioni, C. 1987. *J Immunol* 13, 661–663; Smith, DM et al. 1990. *Cancer Res* 50, 3146–3153.

# Index